历史价值的适应性再利用

Adaptive Reuse with Historical Value

[丹] BIG建筑事务所等 | 编

司炳月 许正阳 姜博文 | 译

大连理工大学出版社

历史价值的适应性再利用

大型办公楼：可持续发展的新定位

历史价值的适应性再利用

大学是公共空间吗?

Major Offices: New Positions in Sustainability

Adaptive Reuse with Historical Value

Universities as Common Space?

始于自然

“先于智慧绽放的美”——因斯布鲁克大学建筑理论系作品展

After Nature

Beauty before Wisdom exhibition showcases the work behind *architecturaltheory.eu*

Bart Lootsma

architecturaltheory.eu不仅仅是几个网站的域名，它也是因斯布鲁克大学建筑理论系的系名。这些网站很好地展示了该系的研究内容，令人印象颇深。“我们网站从文本、书籍、视频和展览四个板块来构建建筑理论。在教学方面，我们不仅开设一系列的讲座和研讨会，还成立了工作室。”

2019—2020年秋冬，为庆贺因斯布鲁克大学建校350周年，其建筑理论系在蒂洛尔州立博物馆举办了一场展览，主题为“先于智慧绽放的美——艺术的知识与科学的艺术”，[1]策展人是克里斯托夫·贝尔茨、罗莎娜·德马特和海伦娜·佩雷尼亚。此次展览将艺术和科学视为互补而又相互竞争的领域，将艺术作品和科技研究成果并列展出，除此之外，还举办了一些讲座和演出。

该系认为“建筑理论”是一门涉及历史、批判和思辨的混合学科。这个主张完全符合策展方的理念，即“艺术与科学是两个相互竞争、相互补充的领域”。在这个激动人心的领域中，“艺术与科学将相互邂逅，相互影响”。策展方认为，对于艺术与自然的关系，人们往往以一种过于简化的方式来评判它（比如，艺术是“跟随在自然之后”的观念——即，艺术模仿自然、源于自然，或从属于自然），事实上，任何有趣的艺术作品都是在探索艺术与自然之间的复杂关系。[2]长期以来，各种理论一直都把建筑学根植于自然之中，这已然成为一个传统。这些理论都是我们在实践中探索和发展得来的。

建筑理论主要研究建筑理论著作史，以及建筑产生的社会、文化和技术方面的实际条件，可以是批判性或建言性的，还可以是为当代和未来的建筑发展提供变革性意见的。比起人类或社会科学的教学，因斯布鲁克大学的工作室采用的教育模式更类似于建筑设计教学。学生们主要从书中学习，并在墙上展示设计布局。学生们每周都会在小型讲座中做一次作品展示，然后就此作品的进展进行公开辩论或批评：可见建筑理论是一种活生生的实践。最后，学生们的作品会发表在architecturaltheory.txt网站上，如果合适的话，也可以在其他地方发表。和理论描述的文档同样重要的是作品的视

architecturaltheory.eu is not just the name of several websites, it is also the name of the Department for Architectural Theory at the University of Innsbruck. The websites give a good impression of what the department does: we produce architectural theory in the form of texts, books, videos and exhibitions and we teach not just lecture series and seminars but also studios.

In Fall/Winter 2019/2020, architecturaltheory.eu is working on an exhibition Schönheit vor Weisheit, Das Wissen der Kunst und die Kunst der Wissenschaft (Beauty before Wisdom, The Knowledge of Art and the Art of Science)[1], at the Ferdinandeum, the Tyrol state museum, Innsbruck. The exhibition is curated by Christoph Bertsch, Rosanna Dematté and Helena Pereña for the 350th anniversary of the University of Innsbruck; it sees art and science as simultaneously complementary and competing fields. The exhibition shows art works alongside products of scientific research and, in addition to that, presents a program with lectures and performances.

The way that architecturaltheory.eu understands “architectural theory”, is as a hybrid discipline of history, criticism and speculation. This fits perfectly with the curators' idea that “art and science are two fields that both compete with and complement each other”, opening up an exciting area “where art and science meet and influence each other”. The curators believe that art is too often judged in a simplistic way with regards to its relationship with nature (as in, the idea that art comes “after nature” – copying it, stemming from it, or subordinate to it), whereas, in reality, it is exactly this complex relationship between art and nature that is explored in any interesting work of art.[2] Architecture also has a long tradition of different theories which root it in nature; these are explored and developed in our work.

Architectural theory investigates the history of architectural treatises, as well as the actual conditions in which architecture is produced, be it social, cultural or technological. Theory criticizes, makes suggestions, and proposes changes for architecture's contemporary and future development. In our studios at the University of Innsbruck, where we apply an educational model that has more in common with teaching architectural design than with the human or social sciences, students mainly work on books and present the layouts on a wall. The students present their work on a weekly basis in short lectures, after which there is a public debate – or critique – about the development of their work: architectural theory is a lived practice. At the end, the students' work is published on the

觉形象的呈现，包括静态效果、动态效果和现在越来越多人在使用的3D打印效果。平面图、剖面图和立面图仍然是描述建筑的主要语言，渲染效果图和动画对创意的激发、理念的广泛传达也有着越来越突出的作用。

为了在博物馆获得更好的展示效果，我们建造了一面展墙，观众可以在墙上追踪到作品的创作进展。墙的一端是内容展示和视频介绍，另一端是一件雕塑作品，把各种具有象征性的元素梦幻般地展示出来，以表现整个历史上大自然对建筑思想的深刻影响：如科林斯式柱头和女像柱、那迦佛像、大象和海龟支撑着地球的雕像——这些都曾经是建筑雕刻装饰的一部分，甚至现在也不时地得到采用。[3]这座雕塑承载着历史上关于建筑与自然之间的众多思考和概念，就如同为建筑学而立的瘟疫纪念柱，我们一方面感谢它们，另一方面又把它们融合在一个巨大的3D打印数字混搭结构之中，并且任其遗落在那里。

的确，比起文字和口头的描述，数字技术越来越利于图像化的呈现。数字概念和图像可以通过JSTOR、academical.edu、谷歌图像搜索、Tumblr和Pinterest等网站在全球高速流通，这就要求我们对新图像和新概念的形成展开不同的思考。谷歌是奥地利这边最主要的搜索引擎，对于突然出现在同一个领域里的"高级"和"低级"文化，它并未做出区分。正是在这个新颖、快速、直观的竞争环境中，"品味"这个在当代一度遭到轻视和抛弃的概念再一次获得了重视。

乔治·阿甘本在其《品味》一书的开场白中写道："与视觉和听觉享有的特权地位相反，西方文化传统把味觉归为感官中最低级的一种，味觉的乐趣把人和动物联系在一起，印象中找不到'任何道德层面的元素'。甚至在黑格尔的美学思想中，味觉也与视觉和听觉这两种'理论上的'的感觉相对立，很明显，一件艺术作品的味道本身是不可以像视觉、听觉那样被品尝的，因为品味一件作品不是独立地、不受约束地去对待它，而是以一种实际可行的方式对其进行处理，使它逐渐地被溶解、被消耗。"[4]连马塞尔·杜尚在几次采访中都表示，"品味，无论好坏，都是艺术最大的敌人。"[5]品味会导致重

website architecturaltheory.txt – and possibly elsewhere if appropriate. Visual images – still, moving and increasingly 3D-printed – play as important a role as text. Plan, section and elevation still form the primary language of architecture, with renderings and animations increasingly important in the generation of ideas, and communicating these ideas to larger audiences.

To accommodate our presence in the museum we constructed a wall, on which viewers can trace the progression of the works in progress. At one end are presentations and videos, and on the other end of the wall there is a sculpture – a hallucinatory dream of figurative elements relating to the ways in which nature has played a role in architectural thinking throughout history: from the Corinthian capital and caryatids, to the Naga Buddha statue, to the image of elephants and a turtle supporting the earth – all of these have been, and sometimes still are, part of architecture's sculpted ornamentation.[3] The sculpture is a kind of architectural Plague Column, in which we express our gratitude to the multitude of thoughts and concepts on the relationship between architecture and nature throughout history, while at the same time merging them and leaving them behind in one big 3D-printed digital mashup.

Indeed, digital technology increasingly favors image production over the written and spoken word. The speed of the global circulation of digital concepts and images, including those available on JSTOR, Academia.edu, Google image search, Tumblr and Pinterest, demands a different kind of reflection on the production of new imagery and concepts. Google, our hegemonic search engine, does not distinguish between "high" and "low" culture, which suddenly appear in an equal field. It is in this new, fast, intuitive playing field that "taste", scornfully dismissed in the Modern epoch, gains importance again.

In the opening words of his essay on taste, Giorgio Agamben states that "contrary to the privileged stature that has been granted to sight and hearing, the Western cultural tradition classifies taste as the lowest of the senses, whose pleasures unite man with other animals, and in whose impressions one will not find 'anything moral'. Even in Hegel's Aesthetics (...) taste is opposed to the two 'theoretical' senses, sight and hearing, since a work of art cannot be tasted as such, because taste does not leave its object free and independent, but deals with it in a really practical way, dissolving and consuming it."[4] Marcel Duchamp, in several interviews, even went so far to say that he considered

复，重复一种风格，重复一种创作方式，这偏离了杜尚对艺术作品所要求的独特而精确的思想内涵。

当然，在建筑学中，重复在其大部分历史时期中都不算是什么问题。重复甚至连一种创作方法都算不上，但它在建筑中未必就没有好处：它使一座新建筑能与一个既定的建筑社区相称，保证整体的连续性。这就是为什么关于风格的争论一直延续至今。如今，建筑师帕特里克·舒马赫更是提出了参数化主义应该会成为下一时期建筑的主导风格这一观点。[6]

通常来说，建筑理论就跟杜尚的主张一样，对品味和当代风格持有怀疑态度，它们在观察和分析是什么驱使当代建筑超越了直接可见的范围。建筑艺术不仅仅是一种视觉现象，它还是一种不断变化的组织形式，这种组织形式必须应对不断变化的社会、文化、政治、经济和技术方面的需求。

然而，品味可能不仅仅是简单的重复：它甚至可能成为创新的必要动力。阿甘本将品味定位为"西方文化构建一种理想的知识体系时所描绘的轮廓，它表现为知识体系最为完备的一种状态，同时，他也强调了获得这种知识是不可能的。这种知识可以把在可感知与可理解之间形而上学的隔阂缝合起来，而其主体事实上并不知道这种知识，因为他们无法解释这种现象"。[7]品味"处于知识和快乐的极限"。它被情感本能所驱动，作为一种"主体不知道但只能渴望的知识"。[8]因此，阿甘本提出品味是哲学的精髓，因为"对知识、哲学的热爱，意味着美必须拯救真理，真理必须拯救美。在这种双重救赎中，知识才能得以实现"。[9]

在设计中，品味作为知识和直觉混合后的奇特产物，在这个世界中变得至关重要，因为整个世界正在被图像主宰，它们正以快得惊人的速度传播着。这些图像可以经过拼贴和组合得到使用，或者，现在甚至以混搭和模型组合的技术使它们彼此互换，制作出新的图像。这两种技术都起源于流行文化。

混搭和其他许多数字创新技术一样，源于音乐领域，比如采样。"混搭是一种创作方式，通常以歌曲的形式出现，由

"taste – bad or good – the greatest enemy of art."[5] Taste leads to repetition – in a style, in a way of working – that distracted from the unique and precise intellectual content Duchamp demanded from a work of art.

Of course, in architecture, repetition has been less of a problem for most of its history. Repetition, not even as a recipe, is not necessarily bad in architecture: it enables a certain continuity in the built environment into which new buildings fit. That is why debates on style have a continuity through the ages. Today, the architect Patrik Schumacher even proposes that parametricism should be the next hegemonic style.[6]

In general, architectural theory, observing and analyzing what drives contemporary architecture beyond the immediately visible, has been as skeptical about taste and contemporary styles as Duchamp was. Architecture is not just a visual phenomenon, it is also a constantly changing form of organization that has to deal with changing social, cultural political, economic and technological demands.

Taste, however, may be much more than simple repetition: it might even be the necessary drive for innovation. Agamben positions taste as "the figure through which Western culture has established an ideal of knowledge, that it presents as the fullest knowledge, at the same time as it underlines the impossibility of attaining such knowledge. Such knowledge, which could suture the metaphysical scission between the sensible and the intelligible, the subject does not in fact know since he cannot explain it."[7] Taste is "situated at the very limit of knowledge and pleasure." It is driven by eroticism, as a "knowledge the subject does not know but can only desire".[8] As such, Agamben even presents taste as the very essence of philosophy, as "love of knowledge, philosophy, signifies that beauty must save truth and truth must save beauty. In this double salvation, knowledge is realized."[9]

In design, taste – this strange hybrid between knowledge and intuition – becomes essential in a world dominated by images circulating at increasingly excessive speeds. These images can be used in collages and assemblages, or now even morph into one another to produce new images in mashups and through kitbashing. Both techniques originate in popular culture.

Mashups originate – as with many digital creative techniques, like sampling – in music. "A mashup (...) is a creative work, usually in a form of a song, created by blending two or more pre-recorded songs, usually by overlaying the vocal track of one song seamlessly over the instrumental track of another."[10] The same can be done with other cultural content, be it images or architecture. Kitbashing, or model bashing, originates in a practice whereby a new scale model is created by taking pieces out of commercial model kits. These pieces may be added to custom projects or to other kits. For professional modelmakers, kitbashing is popular to create concept models for detailing special

两首或多首预先录制的歌曲混合而成，通常是将一首歌曲的人声音轨无缝地覆盖在另一首歌曲的伴奏音轨上。”[10]其他的文化内容也可以如法炮制，不管是图像还是建筑。“模型组合”或“model bashing”都起源于这样一种实践，即通过从商业模型工具包中取出一些片段来创建一个新的比例模型。这些部件也可以添加到自定义项目或其他工具包中。“模型组合”在专业的模型制作者群体中非常流行，常用于制作一些概念模型，给电影特效做细节化处理。商业模型工具包是“细节化处理”的一个现成来源，提供了大量相同的、批量生产的组件，这些组件可以用来向现有的模型添加精微的细节。[11]利用互联网庞大的图像库和3D对象库，“模型组合”也日益成为一种数字现象。这些组件的来源是一些给定的素材，即仿照现实素材进行的建模，所以生成的新图像从一开始就能引发一些联想，具有一定的意义，还能烘托一些气氛，使细节更加令人信服。这些图像在功能、意蕴、历史和叙事上具有一定的暗示作用。使用现有的素材（即，创作始于自然）对于创作至关重要，尤其是科幻小说和玄幻作品（例如，《2001：太空漫游》《星球大战》《银翼杀手》）以及奇幻蒸汽朋克作品（如，《权力的游戏》《狂欢命案》）。另外，还有其他方式的创作，它们是通过在众多图像中出现的小错误、误解和变异形成的，“山寨”这一概念就体现了这种创作形式。[12]

这也意味着没有唯一正确的方法来开发适合计算机和互联网时代的建筑设计。大多数建筑师似乎仍然遵循着戈特弗里德·森佩尔在其著名宣言《科学、工业与艺术》中阐述的方案创作。该宣言于1966年在《新包豪斯》丛书中再次发表，[13]里面提供了一个设计大纲，旨在把工业生产更加恰如其分地描绘出来，去除所有的浮华雕饰，这一点从森佩尔在1851年伦敦世博会上对他看到的一些展品发表的批评中也得以体现。这本书摒弃了数百年来有关建筑设计的理论和既定的知识体系，激发全世界前卫派人士制订新的工艺和工业生产计划。但是，也许创新不仅仅是由新的生产方式引起的，而是由知识和欲望之间的许多裂缝和障碍引发的。也许这是在现代社会重新思考建筑与自然之间关系的一次机会。

effects in movies. Commercial model kits are a ready source of “detailing”, providing any number of identical, mass-produced components that can be used to add fine detail to an existing model.[11] Increasingly, kitbashing is also a digital phenomenon, taking advantage of the vast image and 3D object libraries available on the internet. Because of their origin in given originals, which were modelled after a reality, the new imagery is already charged with associations, meanings and atmospheres from the beginning, which becomes even more convincing in the detailing. It suggests functions, meanings, histories and narratives. The use of existing imagery (made after nature) is essential to the success of immersive effects within science fiction and fantasy productions, like 2001: A Space Odyssey, Star Wars and Blade Runner or fantasy and steam punk productions like Game of Thrones or Carnival Row. Innovation also occurs in other ways, through small mistakes, misinterpretations and variations in the multitude of images produced, as happens in the Chinese concept of innovation, Shanzhai.[12]

This also means there is not one right way to develop architecture and design that fits the age of computing and the internet. Most architects still seem to follow the program as formulated by Gottfried Semper in his famous manifesto Wissenschaft, Industrie und Kunst, as republished in the series Neue Bauhausbücher in 1966.[13] This book was a program for a design that would do more justice to industrial production and get rid of all ornamentation, which Semper had criticized in the objects he found at the World Exhibition in London in 1851. It triggered avant-gardes worldwide to formulate new programs for craft and industrial production, abandoning centuries of theory and knowledge about architecture and design. But maybe innovation is not only triggered by new ways of production, but in the many cracks and fissures between knowledge and desire. And maybe this is an opportunity to rethink the relationship between architecture and nature – after nature.

1. Design team: Bart Lootsma, Giacomo Pala, Stefan Maier, in collaboration with David Kienpointner and Prashant Chavant
2. http://www.tiroler-landesmuseen.at/page.cfm?vpath=programm/ausstellungen&snippetmode=future&genericpageid=10565
3. Giacomo Pala. *Hypnerotomachia Naturae*, http://www.architecturaltheory.eu/architekturtheorieINFO/wp-content/uploads/2019/09/Giacomo-Pala_Hypnerotomachia-Naturae_Brochure-Version_20190923.pdf
4. Giorgio Agamben. *Taste*, (London/New York/Calcutta: Seagull Books, 2017.)
5. See ao.: https://www.quora.com/What-did-Duchamp-mean-by-Taste-is-the-enemy-of-art
6. Patrik Schumacher. *Hegemonic Parametricism delivers a Market-based Urban Order*, in: H. Castle (ed.), Patrik Schumacher, (ed.), AD Parametricism 2.0 - Rethinking Architecture's Agenda for the 21st Century, AD Profile #240, March/April 2016.
7. See note 5. p. 51. / 8. Idem. p. 52. / 9. Idem. p. 76. /
10. https://en.wikipedia.org/wiki/Mashup_(music) / 11. https://en.wikipedia.org/wiki/Kitbashing / 12. Byung-Chul Han. *Shanzhai*, (Berlin: Merve Verlag, 2011.)
13. Gottfried Semper. *Wissenschaft, Industrie und Kunst*, (Mainz/Berlin: Neue Bauhausbücher, Florian Kupferberg, 1966.)

大型办公楼：可持续发展的新定位

Major

New Positions i

一座总部大楼管理着一个机构的所有运作，因此它的最主要功能是办公。但是，其设计也可以服务于其他功能，例如，表现公司的价值观。当代的企业环境在很大程度上起源于办公楼的建筑历史，并已经朝着提高效率和员工福利的方向发展。

如果土地价格低于中央商务区的价格，那么总部可以分散建立。一个机构可以在他们的私人土地上建立一个园区。在郊区或者城市边缘地带，离开了城市环境，建筑师可以从形式和外观的种种

A headquarters (HQ) building governs the operations of an organisation, and so it is primarily an office building. However, the design may serve other functions, such as representing company values. The contemporary corporate environment owes much to the architectural history of offices, which has trended towards efficiency and enhancing the well-being of employees.
Where land prices are lower than in a central business district (CBD), the HQ can spread out. An organisation may be able to build a campus on their private territory. Without urban context, the

限制中摆脱出来。但是，这些地带对建筑的可持续性依然会产生影响。

我们考虑了四座建在CBD以外的总部大楼和园区的案例。我们考虑到当前处于"气候紧急状态"这一实际情况，发现了建筑设计中一些涉及可持续性的重要问题，特别是一些可以支持汽车停放和通行的办公楼设计。

architect is freed of constraints on form and surface. But sustainability is effected by locations that are suburban or on the urban edge.
We consider four HQ buildings and campuses beyond the CBD. Acknowledging the climate emergency, we identify major overlooked issues of sustainability, particularly building offices that encourage car usage.

大型办公楼：可持续发展的新定位
Major Offices: New Positions in Sustainability

Herbert Wright

大型办公项目的建筑设计是由许多因素决定的，首先是它的地理位置。过去，办公楼多聚集在中心城市，成为密密麻麻的城市结构网的一部分。高昂的地价助长了高容积率（建筑面积与场地面积的比率）。在密苏里州圣路易斯市中心，路易斯·沙利文和丹克马尔·阿德勒设计了世界上第一座钢结构摩天大楼——温莱特大厦（1891年），作为酿酒协会的总部。但随着汽车的兴起，地价较低的郊区或城市边缘地区成为开发写字楼的诱人地带。场地的扩大使得这些机构可以将一批建筑聚集到经过景观园林绿化的各个园区中。建筑设计多为低层或中层，由于没有相邻的建筑和城市街道需要协调，因此建筑师可以更加自由地进行设计表现。我们经过考察，考虑了四个这样的总部和园区的项目。虽然可持续性已被纳入建筑议程中，但还有一个重要因素被低估了，那就是这个地点的流动能力问题。

一座重要的办公大楼能够反映其入驻机构的一些内涵。它可以实际折射出企业价值和品牌力量。例如，福斯特合伙人事务所在加州库比蒂诺市为苹果公司设计的1.6km宽的环形园区，作为苹果公司总部（2017年）。它就像苹果公司一样，描绘了一个大规模的未来主义愿景。

建筑师弗兰克·劳埃德·赖特为提高工人的舒适感，开创了内部建筑的基本理念。第一座开放式办公楼——拉金大厦于1905年诞生于纽约州的布法罗市。这是一家香皂公司的总部，总部大楼的上方有一个中庭。位于威斯康星州瑞辛市的庄臣公司总部（1939年）有一个令人叹为观止的开放式“大工作室”，里面外形优雅的柱子向天花板的方向逐渐变宽。办公建筑还在继续发展。从1958年起，“办公景观”的概念打破了一人固定一张办公桌的布局，引入了盆栽。1996年提出的ABW（移动式办公）模式让员工可以自由改变和选择他们的办公桌（即，轮用办公桌）。此时的建筑设计也响应了这个理念，并如我们的调查

The architecture of a major office project is shaped by many factors, starting with its location. Once, offices clustered in the central city and were part of their tight urban fabric. High land prices encouraged a high plot ratio (the ratio of floor area to site area). In downtown St Louis, Missouri, Louis Sullivan and Dankmar Adlar designed the first steel-framed skyscraper, the Wainwright Building (1891), a brewery association HQ. But suburban or urban edge territory, where land prices are lower, made attractive sites for office development with the rise of the automobile. Larger sites allowed organisations to group buildings into campuses which are landscaped. The architecture tends to be low- or medium-rise, and without neighbouring buildings and city streets to respond to, the architect has more freedom of expression. In our survey, we consider four such HQ and campus projects. While sustainability is embedded in architecture's agenda, a vital but undervalued factor is the mobility issues of location, which we shall return to.

A major office building says something about the organisation occupying it. It can act as a physical manifestation of company values, or a projection of brand power. For example, the 1.6km-wide ring-shaped Apple Park HQ (2017) by Foster+Partners in Cupertino, California, is a futuristic vision on a vast scale, just like the Apple company.

Frank Lloyd Wright pioneered the basics of internal architecture that aimed to enhance the well-being of workers. The first open-plan office floor was at the Larkin Building (1905) in Buffalo, New York, a soap company HQ. Above it was an atrium. Johnson Wax Headquarters (1939) in Racine, Wisconsin has a breath-taking open-plan "Great Workroom", with elegant columns which widen towards the ceiling. Offices have continued to evolve. For example, from 1958, the Bürolandschaft (office landscape) concept broke up regimented desk layouts and introduced pot-plants, and in 1996 Activity Based Working allowed workers to change and choose their desk

温莱特大厦，路易斯·沙利文和丹克马尔·阿德勒，密苏里州圣路易斯，1891年
Wainwright Building by Louis Sullivan and Dankmar Adler in St. Louis, Missouri, 1891

庄臣公司总部，弗兰克·劳埃德·赖特，威斯康星州瑞辛市，1939年
Johnson Wax Headquarters by Frank Lloyd Wright in Racine, Wisconsin, 1939

即将显示的那样，这种为办公环境而产生的创新还在继续。

我们第一个要说的是瑞士洛桑市一个公园的湖滨项目。该项目名为奥林匹克之家（16页），是国际奥委会的新总部大楼，占地22000m²，由总部位于哥本哈根的3XN建筑事务所设计。告别腐败、使用兴奋剂以及给奥运会主办城市带来巨额债务的历史，国际奥委会重新整合为一座大楼，以此为契机，透过大楼的设计展示他们所维护的价值观，包括公开透明、和平、协作精神和致力于可持续发展。建筑一层的屋顶是一个花园露台，像是对公园的延伸。上方的三个办公楼层外部包裹了线条如流水般的双层玻璃幕墙，形成了一个弯曲的四臂星形平面，在造型上象征着和平鸽。玻璃外侧的竖向构件在波浪中轻轻弯曲，暗示了运动性。要说这是对运动中的运动员们的一种模拟，其实有些微妙，但比象征鸽子的说法更有说服力。立面通透感十足的新总部大楼与旁边前总部传统封闭的城堡型建筑形成了鲜明的对比，似乎是对该组织发生变革的一种隐喻。

连续的楼板横贯整座建筑，打造出许多阳光充足的开放式办公空间和休息空间，但姿态更为突出的要属中央楼梯。自从3XN建筑事务所在哥本哈根的欧瑞斯塔学院（2007年）的中心设置了一个巨大的浮动螺旋楼梯后，这种做法就一直沿用下来，以宣传大型开放式楼梯对社交的促进作用。3XN的负责人金·尼尔森称这种楼梯为"催化剂"，因为人们可以在上面会面，而且可以在视觉上与地板上的任何地方产生联系。它与国际奥委会的合作精神产生了共鸣。这些楼梯重新诠释了奥林匹克五环标志，因为每层楼都是由上而下的螺旋形楼梯组成的。你仍然可以从正中间往下看，从下面看，木质楼梯就像向你滚来的圆环。楼梯上方是一扇圆形的采光天窗。

即使不是出于国际奥委会的价值理念，3XN建筑事务所也会保证建筑具有良好的可持续性。奥林匹克之家几乎回收利用

("hot desking"). Architecture responded, and our survey will show that it continues to innovate for the office environment.

We start with a lakeside project in a park in Lausanne, Switzerland. Olympic House (p.16) is the new 22,000m² HQ of the International Olympic Committee (IOC), designed by Copenhagen-based practise 3XN. Moving on from the history of corruption, doping and leaving Olympic Games host cities with colossal debts, the IOC's consolidation into one building gave a chance to assert their values, which include transparency, peace, collaborative spirit and commitment to sustainability, in the architecture. There is a garden roof terrace over the ground floor, like an extension of the park. The three office floors above have double-glazed facades which flow around the building, making a curvy four-armed star shape plan which is an abstraction of the dove of peace. Upright elements outside of the glazing gently bend from the vertical in waves, suggesting movement. This reference to athletes in action is subtle, but more convincing than the dove. The visible contrast of the transparency facades with the opacity of the traditional chateau next to it, which was the previous IOC HQ, is like a metaphor for change in the organisation.

Continuous floorplates across the building create light-filled open-plan offices and plenty of break-out space, but the stand-out architectural gesture is the central staircase. Ever since 3XN placed a grand floating spiral staircase at the heart of Ørestad College (2007) in Copenhagen, the practice has evangelised the social role of dramatic open staircases. 3XN principal Kim Nielsen calls them "catalysers" because people meet on them and can connect visually with anywhere on the floors. It resonates with the IOC's spirit of co-operation. Their stairs re-interpret the Olympics' five-ring logo, because on each floor, they spiral within a circle displaced from those above or below. You can still look right down the middle, and from below, the wooden staircases look like rings tumbling towards

苹果公司，福斯特合伙人事务所，加州库比蒂诺市，2017年
Apple Park HQ by Foster + Partners in Cupertino, California, 2017

了前一栋建筑的全部材料，并在屋顶安装了光伏发电装置，与湖水的热交换装置和雨水回收装置。

一家运动服装品牌总部的规模可能是国际奥委会总部的两倍之多，而员工数量可能是其四倍，这说明了体育运动商业化所产生的影响。德国斯图加特市的Behnish Architekten建筑师事务所给阿迪达斯集团设计了一座总面积为52000m²的"体育世界"（34页），它并非真的是一座体育场，这座办公大楼背后也有一段有趣的故事。阿迪达斯及其竞争对手彪马两家公司的创始人是兄弟关系，二人相互竞争，且两家公司的总部仍然都位于德国小镇黑措根奥拉赫。彪马总部（2009年）由克劳斯·克雷克斯设计，是一座三层高的红色矩形建筑，位于一座小园区内，与阿迪达斯"体育世界"大楼相距1km。而阿迪达斯的这座总部大楼也是一栋三层的矩形办公楼，似乎有意表达了与彪马的竞争关系。

阿迪达斯"体育世界"办公大楼外观为矩形，长143m、宽118m，不仅规模更大，而且上面两层被高高地从地面上托起。员工和访客通过下方一层的入口进入一个宽敞开阔、楼角分明的大厅。视频屏幕传达着阿迪达斯的品牌信息。大楼的天花板通向一个玻璃中庭，中庭像是大楼的颈部，通过一个露天的夹层向上延伸到办公楼层。沿着黑色的线性楼梯和与墙面成一定倾角的过道可以通向办公楼层。混凝土墙面、灰黑的色调以及中庭的安排，既有工业化的感觉，又有现代化的氛围，而建筑棱角分明的特征颇有些结构主义的味道。

相比内部空间，"体育世界"悬浮的体量是直线形的。建筑立面被包裹在醒目的金属网格中，可以有效地调节日光。这种盒子式的结构由空腹桁架构成，它就像没有斜梁的水平梯子。在建筑内部，中庭继续向上延伸，每层都有一条宽阔的轴向通道或"主干道"。中庭提供了沿着楼层和夹在楼层之间的视觉连接，但从这里看向那一排排的门时，感觉和监狱的景象没什么两

you. Above the stairs is a circular skylight.

3XN would have built in good sustainability even it had not been an IOC value. Olympic House recycled almost all of a previous building, and has roof photovoltaics, heat exchange with the lake water and rainwater re-use.

It says something about the commercialisation of sport that a sportswear brand HQ can be over twice the size of the IOC's, and have four times as many workers. The 52,000m² Adidas World of Sports ARENA (p.34), designed by Stuttgart-based Behnish Architekten, is not an arena. It also has an interesting backstory. Adidas and its rival Puma were each set up by competing brothers and are still both based in the small German town of Herzogenaurach. A kilometre away from the ARENA, the Puma Vision Headquarters (2009) is a bright red three-storey rectangular building in a small campus, designed by Klaus Krex. The rivalry seems to find expression in a three-storey rectangular office building for Adidas.

143m by 118m in area, the rectangular ARENA office box is not only bigger, but is raised two storeys above ground. Beneath it, a ground floor entrance welcomes visitors and employees into a wide angular hall. Video screens convey Adidas branding. Its ceiling opens up into a glazed atrium, which rises like a neck through an open-to-air gap level to the office floors above. They are reached by black linear staircases and gangways angled from the walls. The concrete surfaces, shades of grey and black, and atrium feel simultaneously industrial and contemporary, with a touch of deconstructivism in its angularity.

The ARENA's floating volume, by contrast, is rectilinear. The facades are wrapped in a striking lattice which actively regulates daylight, and the box structure is made by Viereendel trusses, which are like horizontal ladders without sloping beams. Inside, the atrium continues upwards, punching through a wide axial passage or "main street" on each floor. It provides visual connectivity along and between floors, but some views across it to rows of doors

瑞士奥利匹克之家
Olympic House, Switzerland

阿迪达斯"体育世界"办公大楼，德国
Adidas World of Sports ARENA, Germany

样。工作区域位于楼板的边缘地带。办公桌是正交排列的，好像没有一点儿办公室的感觉。它们中间点缀着五颜六色的金属"集装箱"，这些箱式空间可以用作会议室或私人空间。即使如此非正式的元素，也与整个矩形平面保持精确的对齐。

现在我们从各个公司的总部来到他们设立具体战略部门的园区。在我们的调查中，土耳其ERA建筑师事务所设计的加兰蒂银行BBVA科技园区（52页）所在的地段都市化气息最为浓厚。它虽与伊斯坦布尔市中心相距30km，但仍属于这座拥有1500万人口的大都市范围之内。高速公路穿过密集的居民区，为物流和大卖场提供了一种"边缘城市"环境。这个园区是道路系统中的一个孤岛，从这里望去，它就像一座棱角分明的堡垒，外面包裹着散发现代主义气息的玻璃幕墙。事实上，这座四层办公楼"悬浮"在一个极其别致的城市生活区之上，生活区由一些支撑性的裙房构成，还做了景观设计，人工山丘就是其中一项。在BBVA园区，有两个浮动体量，一个折叠成"e"形，另一个呈"H"形。一系列露天空间由办公室和连接它们之间的桥梁通道所包围。建筑整个外围包裹着一层连续玻璃幕墙，隔开了屋顶的空气调节系统，和内侧形成双层玻璃立面，还把各个建筑开口都包裹起来。整个玻璃立面把建筑体量围绕了起来，形成了一个绵长统一、棱角分明的建筑体块。它的这种"悬浮"形式与阿迪达斯总部的建筑案例几乎没有什么共同之处，但与中国地产公司深圳总部万科中心（2009年）却有几分相似。万科中心是由斯蒂文·霍尔建筑师事务所设计的超大型建筑，一个悬浮的线性体量在做过景观设计的地面上形成了几个折角。

BBVA园区拥有53500m²的高档开放式办公室。在交通流线区域可以看到挑空空间，这些空间比庭院还大，因为它们通向下方连续的空间。值得注意的是，办公大楼的角落在外部是尖锐的，但在内部是圆滑的，这有助于园区内部形成有机的感觉。支撑办公空间的混凝土核心筒是圆形的。道路穿过景观，延伸至场地边界，容纳了浮动体量的入口。和万科中心的设计一样，

are not unlike in a prison. The work zones lie around the periphery of the floorplates. Desks are organised in orthogonal arrays as if Bürolandschaft never happened. They are interspersed with colourful metal boxes like shipping containers for meetings or privacy. Even these informal elements are aligned precisely with the whole rectangular plan.

We now go from HQs to campuses where corporations have headquartered specific strategic divisions. In our survey, the location of the Garanti BBVA Bank Technology Campus (p.52) by Turkish practice ERA Architects is the most urban. It is 30km from the centre of Istanbul but within its conurbation of 15 million. Expressways slice through dense residential areas and feed an "edge city" environment of logistics and big box retail. This campus is an isolated island in the road system, from which it looks like an angular fortress of modernist glass curtain walls. In fact, the four-storey office building floats above an extraordinary village of supporting podium buildings and landscaping, including artificial hills. At BBVA, there are two floating volumes, one folding to create an "e" shape and the other an "H" shape. A chain of open-air voids are enclosed by the offices and connecting bridge passages between them. A continuous peripheral glass facade, which also screens the HVAC on the roofs, provides double glazing but also glass screens across building gaps. It wraps around the volumes to create a long unified angular block. The floating form has little in common with ARENA, but has some similarities to the super-scaled Steven Holl Architects-designed Vanke Center (2009) in Shenzhen, a Chinese property company HQ, in which a floating linear volume makes angular turns above landscaped ground.

BBVA's campus has 53,500m² of high-grade open-plan office. The circulation areas look into the voids, which are more than courtyards because they open into continuous space below. Significantly, the office block corners are sharp on the exterior but curved inside, contributing to the organic feel of the campus interior. Concrete

加兰蒂BBVA银行科技园区，土耳其
Garanti Bank BBVA Technology Campus, Turkey

场地中的很多建筑都设计了像山丘一样的绿色屋顶。办公层下面所有的体量（包括一个16000m²的礼堂）都有弯曲的步行道和双层高的大厅。一层成为宁静的避风港，多处空间做了留白处理，彰显了这座建筑的魅力。然而，这个园区对汽车具有高度的依赖性。

另一个在城市边缘地带的案例是印度的电子城，距离班加罗尔市中心约15km。公路、高压线铁塔和规格不一、零散的开发项目是这个城市的一大特点，另一个特点是这个城市遍布着众多的湖泊。塔塔一体化办公园区（72页）就坐落在其中一个湖畔，它由本地建筑设计公司Mindspace Architects建筑师事务所打造，占地36000m²。它本质上是一座笔直而细长的三层建筑，但它的另一边则形成了一种流动而有机的形态，像是展开的双翼。人工湖两侧的建筑呈梯台式上升，形状曲折、面积宽阔，上面栽种着许多树木，形成了花园式露台。

园区边上的一条马路将这个新的人工"生态"湖与当地的天然湖泊分隔开来，尽管如此，天然湖泊仍然给园区的设计带来了灵感和参考。办公楼层下方的一座翼楼里容纳了一间大型餐厅，而另一座翼楼里则是印度塔塔公司的手表部门，主要生产一些工程消费品和珠宝首饰。花园露台可以吸引职员到室外办公。

这座园区在设计上强烈地呼应了葡萄牙里斯本市的尚帕利莫中心（2012年），后者是由印度杰出的建筑大师查尔斯·柯里亚设计的科研医疗中心。两座建筑都是借助低矮弯曲的雕塑性石材构件、室外的小路和石阶，以及建筑和水的互动，创造了一个宁静而超然的环境。但是塔塔办公园区还有其他的元素来营造这种宁静。除了露台和绿色的墙壁，最著名的就是贯穿整座建筑的五个中庭，其形状像从顶部切下的圆锥体，方便采光。漂浮的螺旋楼梯在里面拔地而起。在两翼中间是一个饰有通高玻璃的两层悬浮体块凌驾于水面上，它被嵌在一个体积不大的扇贝状未来主义风格的结构中，看起来稍显突兀。该园区采用

cores serving the offices are rounded. Paths run through landscaping, which extends to the site boundary and hosts entrances to the floating volumes. As at the Vanke Center, some of them are green-roofed forms like hills. All volumes below the offices, including a 16,000m² auditorium, have curved footprints and lobbies are double-height. The peaceful haven of ground territory and inspiring voids are beautiful. But the campus is highly car-dependent.

Another urban edge environment is Electronics City, about 15km from the centre of Bangalore, India. Highways, pylons, and random stand-alone developments characterise it, but it is also dotted with lakes. Beside one is the 36,000m² Titan Integrity Campus (p.72) by local based Mindspace Architects. It is essentially a long straight three-storey building on one side, but on its other side, it is a flowing, organic form, from which two wings extend. They ascend in wide, curving garden terraces planted with trees, either side of a new body of water.

A road divides this new "bio lake" from the natural lake, which nevertheless inspires and informs the campus design. One wing includes a vast dining hall beneath office floors, the other hosts the watches division of Titan India, which makes engineered consumer products and jewellery. The garden terraces invite office workers to work outside.

This campus has a strong echo of the Champalimaud Center, Lisbon (2012), a medical research and office complex designed by the great Indian architect Charles Correa. Both create a transcendental serenity with their low, curving sculptural stone forms, exterior paths and stone stairs, and interaction with water. But Titan Integrity Campus has additional elements building this tranquillity. As well as the terraces and green walls, the most notable are five atria penetrating the building, shaped like cones sliced off at the top to bring light in. Floating spiral staircases rise in them. Between the wings, there is a floating volume – two floors with full-height glazing

塔塔一体化办公园区，印度班加罗尔
Titan Integrity Campus, Bangalore, India

尚帕利莫中心，查尔斯·柯里亚，葡萄牙里斯本，2012年
Champalimaud Foundation by Charles Correa, Lisbon, Portugal, 2012

的可持续发展手段之一就是利用光电技术供应了四分之一的用电。在我们调查的所有建筑中，塔塔一体化办公园区给人的感觉最不像一台机器，它没有将自己隐蔽起来，而是将整个环境都向外界打开，是与自然融合得最为成功的园区。

某些元素在我们的调查中重复出现，如中庭和大楼梯。BBVA园区也设有开放的挑空空间，虽然并非中庭，但也达到了视觉上的连通性。所有建筑都具有可持续发展的特点，奥林匹克之家被评为能源与环境设计先锋白金奖建筑。但是做到比传统办公楼节省三分之一的能源真的就足够了吗？建筑性能需要更接近于零能耗，为此建筑界必须付出更多的努力。

在市中心以外建立办公室是否理所应当呢？城市扩张会侵吞土地。即使是棕色用地（已开发但如今闲置的地带）也可以回归自然，而不是重新开发。非中央商务区的开发很可能要依赖汽车，这是最不可持续的交通工具。"碳简报"网站的数据显示，电动汽车并不能消除温室气体排放，在其使用寿命内，电动汽车的排放量仍然达到了欧盟汽油车平均排放量的50%。为汽车设计的绿色建筑不是真正的绿色建筑。以苹果园区总部为例——尽管它的园区绿化率高达80%，但据报道，它有14000个停车位，与员工人数持平。有的停车场比办公区还多，像比苹果办公园区稍小的加兰蒂BBVA园区，它的地下停车场面积达72500m²。阿迪达斯"体育世界"大楼只是这个园区几座办公大楼其中的一座，该园区一共才有5600名员工，而阿迪达斯的办公楼自己就安设了3500个停车位。奥林匹克之家拥有500多名员工，3XN建筑事务所不仅为它设置了135个自行车停车位，而且在800m外还建有一座地铁站，但该建筑仍有80多个机动车位。在我们的调查中，唯一没有停车场的就是塔塔一体化办公园区。

在非市中心区域建造办公大楼或许会产生令人激动不已的设计，打造出绝妙的办公环境，同时兼具一定的可持续性，可是这种建筑从根本上来说是不可持续的，问题的根源就在于它们的地理位置。

mounted above the water in a small scallop-shaped futuristic volume that is slightly incongruous. Sustainable technologies include photovoltaics supplying around a quarter of power. Of all the buildings in our survey, the Titan Integrity Campus feels the least like a machine. Its environment opens out on the world rather than screens it away. It is the most integrated with nature.

Certain elements have repeated in our survey, such as the atrium and grand staircase. The BBVA has open voids rather than atriums, but they too create visual connectivity. All buildings have sustainable features and Olympic House is LEED-platinum rated. But when energy savings compared to conventional offices may reach perhaps a third, is it really enough? Building performance needs to move much closer to net zero energy. Architecture must try even harder.

Should offices outside of city centers be built at all? Urban sprawl eats land. Even brownfield sites could be returned to nature rather than redeveloped. Non-CBD developments are likely to depend on cars, the least sustainable transport. Electric cars do not eliminate greenhouse gas emissions – over their lifetime they still reach half of an average EU petrol car, according to CarbonBrief. A green building is not green when designed for cars. Consider the Apple Park headquarters – despite its campus being 80% green landscaping, it has a reported 14,000 parking spaces, the same as its staff capacity. There is more car park than office area, just as the smaller Garanti BBVA campus does, with 72,500m² of underground car park. The Adidas ARENA is just one of several buildings in a campus for 5,600 staff, and there are 3,500 car park spaces. At Olympic House with 500 staff, 3XN built storage for 135 bicycles, and a Metro station is 800m away, but the building still has over 80 car parking spaces. In our survey, only the Titan Integrity Campus includes no car park.

Non-central offices may produce exciting designs, great office environments and have sustainable features, but in our climate emergency, most are simply not sustainable because of their location.

奥林匹克之家——国际奥委会总部
Olympic House – IOC Headquarters

3XN

奥林匹克之家在建筑中传达奥林匹克精神和体育运动的核心

Olympic House expresses the core of the Olympic spirit and the movement of sports in architecture

奥林匹克之家的目标是将国际奥委会的工作人员聚集在一起。目前有500名员工分布在洛桑的四个地方。3XN是通过一次分为多个阶段的国际建筑设计竞赛入选的，该竞赛由国际建筑师协会认证，其评委会由著名建筑师组成，是为了联合总部的设计而举办的。

奥林匹克之家是围绕五个关键目标形成的：运动、透明度、灵活性、可持续性和合作，每一个目标都将奥林匹克运动的核心原则转化成为建筑形式。这座建筑真实地反映了奥林匹克主义、奥林匹克运动以及国际奥委会促进合作的功能。

"奥林匹克之家"位于瑞士日内瓦湖附近的一个公共公园内，这里是18世纪维迪城堡之乡。3XN负责该项目建设，旨在实现与独特的自然和历史背景的最高水平的融合，同时为当地社区创造一个标志性的建筑地标。该设计既尊重了城堡的传统，又尊重了公园的环境，在绿色公共空间和奥林匹克之家之间建立了自然的过渡。

3XN设计的一个标志是立面，通过模仿运动员优美的动作来向奥林匹克精神致敬。动态的、起伏的流动型立面从各个角度看起来都是各不一样的，传递出运动员在运动中蕴含的能量。在体育中，运动会带来最佳的表现；同样，建筑外围护结构在形式上的操作也会直接影响其功能。

参考了奥林匹克五环的统一楼梯，从地面延伸到顶部，并通过一个中央中庭连接五个楼层。橡木楼梯遵循主动式设计原则，圈定了社会活动和运动的中心区域，提升了社区感。展览空间、自助餐厅和会议室也布置在中央楼梯的周围，为大楼的使用者提升了社区感。

透明度成为奥林匹克运动具有责任性的一个重要象征，同时也体现了国际奥委会对尊重、友谊和卓越的承诺。

奥林匹克之家符合当地和国际上最严格的可持续发展标准。它被授予LEED白金级别——这是国际LEED绿色建筑项目的最高认证级别；还被授予了瑞士可持续建筑标准和瑞士节能建筑标准中的白金级别。在提高能源和节水效率、减少废物和整合景观方面做出了特别的努力。在不影响工作空间质量的前提下，将建筑的生态足迹降至最低的创新特点，体现了对可持续性的承诺。

该建筑的外层使用了气密性材料和外墙内表面的三层玻璃，起到了出色的保温效果。除了雨水收集和太阳能电池板之外，另一个重要的可持续性特征是通过热交换利用湖水对建筑物进行供暖和制冷。

这种对可持续性的坚定承诺也反映在建设过程中。奥林匹克之家是施工循环经济的典范：原行政大楼建筑材料的95%得到了重复利用或回收。

Olympic House aims to bring together the IOC staff – 500 employees currently spread across four locations in Lausanne – under one roof. 3XN was selected through a multi-stage, international architecture competition, certified by the International Union of Architects and led by a jury of renowned architects, for the design of the consolidated head office.

Olympic House is formed around five key objectives: movement, transparency, flexibility, sustainability, and collaboration, each of which translates the Olympic Movement's core principles into built form. The building authentically reflects Olympism, the Olympic Movement and the role of the IOC as a catalyst for collaboration.

The site is located in a public park, home to the eighteenth-century castle Château de Vidy, near Lake Geneva. 3XN approached the project with the intent to achieve the highest level of integration with the unique natural and historical setting, while creating an emblematic architectural landmark for the local community. The design respects both the Château's legacy and the park setting, establishing seamless transitions between the green public space and Olympic House.

A hallmark of 3XN's design, the facade, pays tribute to the Olympic spirit by emulating the graceful movements of an

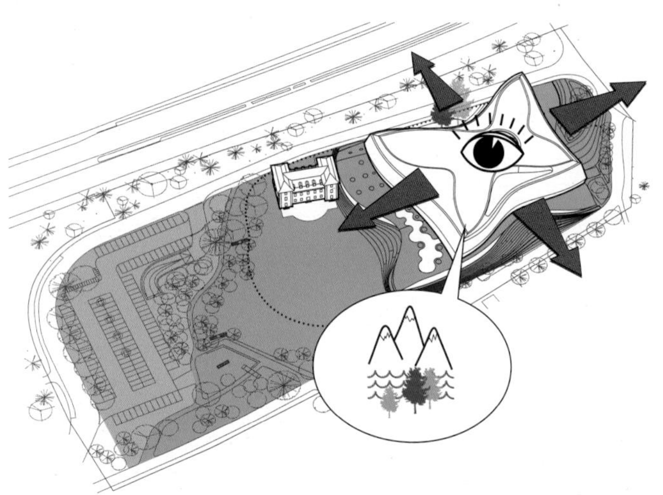

从奥林匹克之家的各个楼层都能欣赏到湖水、公园和城市风光
lake, park and city views from all floors of the Olympic House

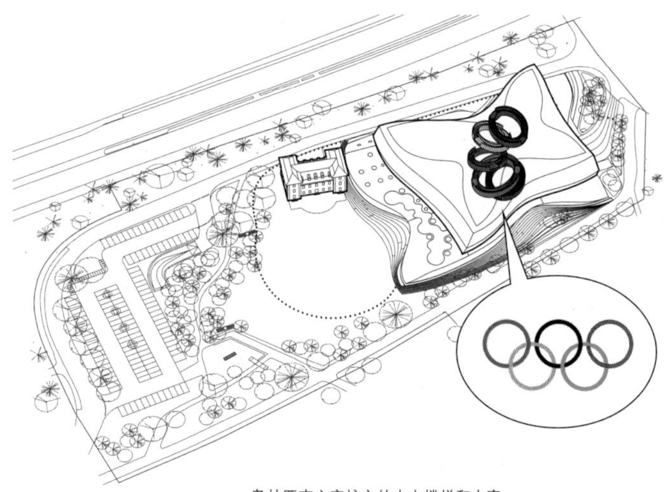

奥林匹克之家核心的中央楼梯和中庭
central stair and atrium at the heart of the Olympic House

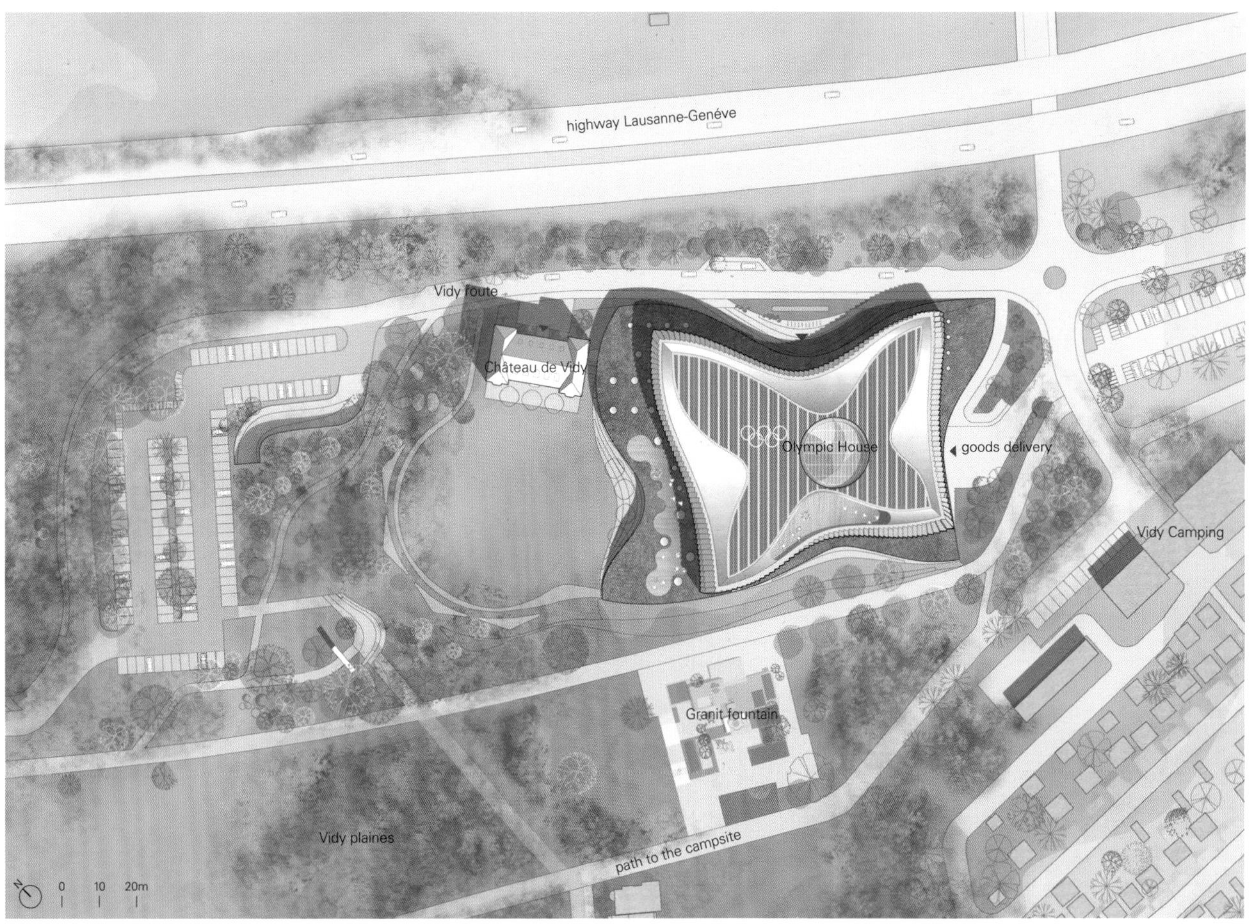
highway Lausanne-Genéve
Vidy route
Château de Vidy
Olympic House
goods delivery
Vidy Camping
Granit fountain
Vidy plaines
path to the campsite
0 10 20m

Hors des cases
des 2 côtés
30

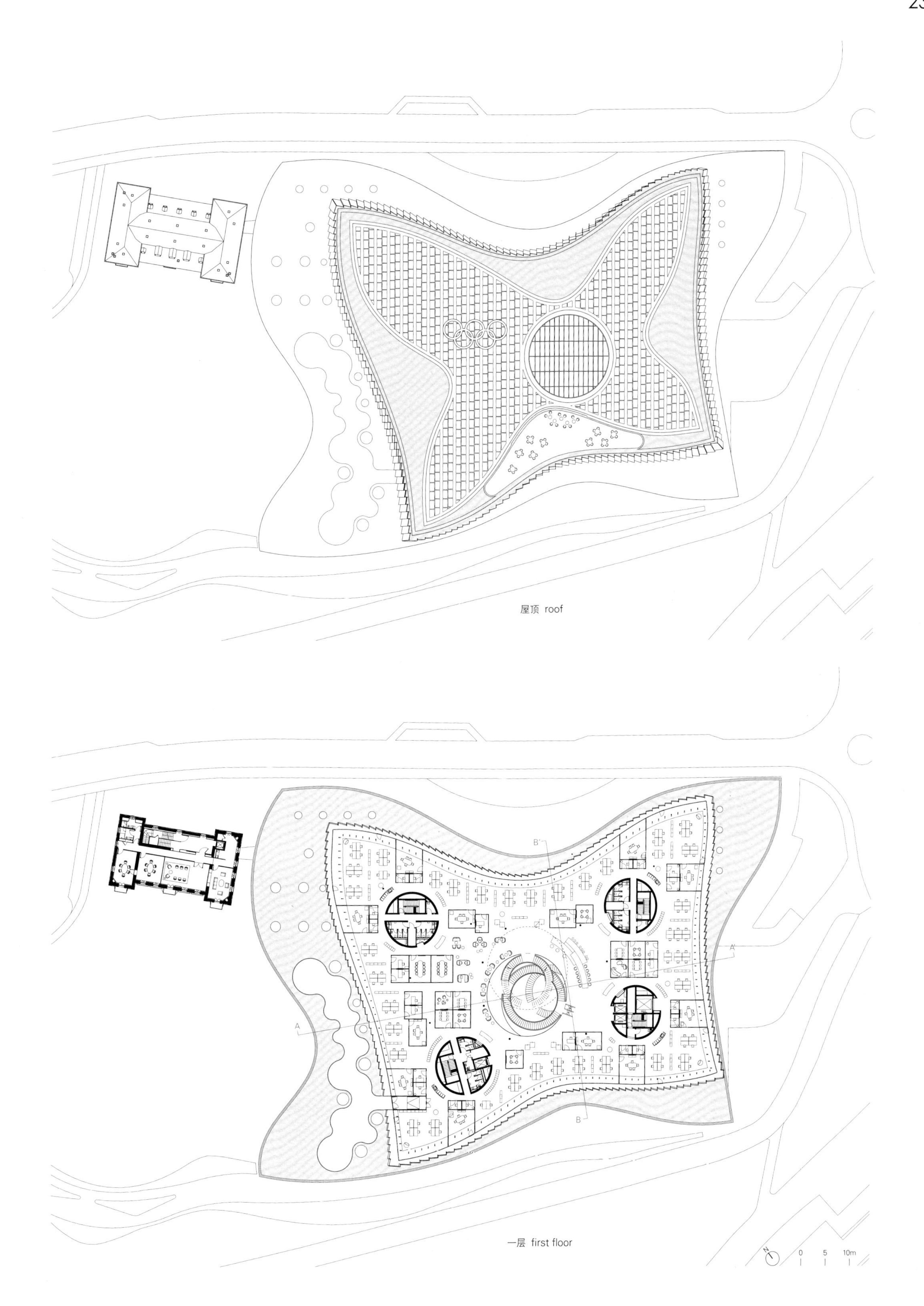

屋顶 roof

一层 first floor

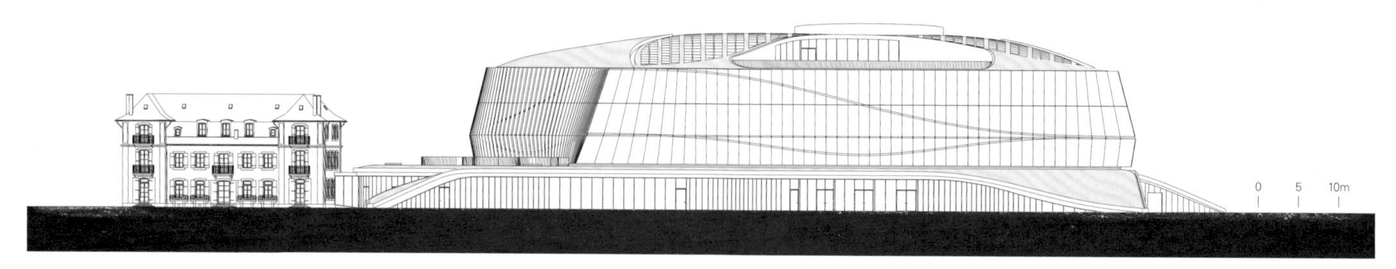

西南立面 south-west elevation

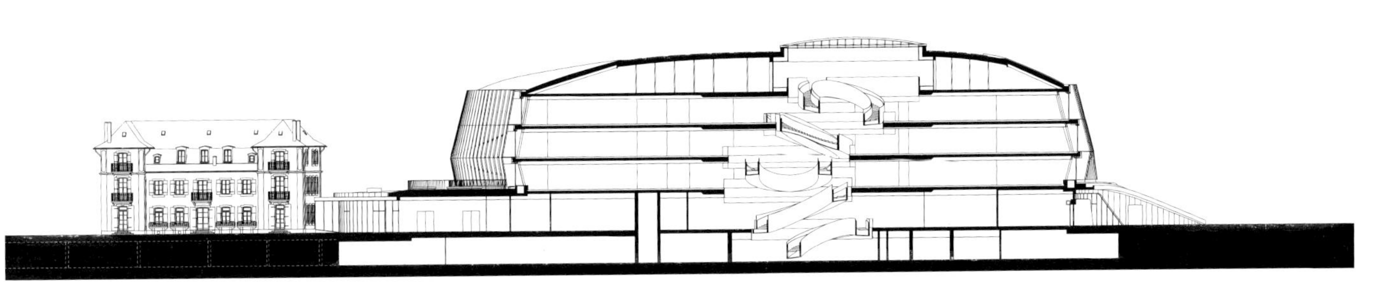

A–A' 剖面图 section A-A'

athlete. The dynamic, undulating flow of the facade appears different from all angles, conveying the energy of an athlete in motion. In sports, movement leads to optimized performance; likewise, the formal manipulations of the building's envelope have a direct effect on how it functions.

The Unity Staircase, which references the Olympic Rings, soars the full height of the building and connects the five floors through a central atrium. Following the principles of active design, the oak staircase defines the central area for social activity and movement, promoting a sense of community. Exhibition spaces, a cafeteria and meeting rooms are also arranged around the central staircase, promoting a sense of community for the building's users.

Transparency becomes an important metaphor for the Olympic Movement's responsibility, and the IOC's commitment to respect, friendship, and excellence.

Olympic House meets the most demanding sustainability standards, both locally and internationally. It was awarded LEED Platinum, the highest certification level of the international LEED green building program, Platinum level in the Swiss Sustainable Construction Standard (SNBS) and the Swiss standard for energy-efficient buildings, Minergie P. Special effort was put into energy and water efficiency, waste reduction and landscape integration. Innovative features that minimize the building's environmental footprint, without compromising the quality of the workspace, demonstrate commitment to sustainability.

The building's envelope allows excellent insulation through airtightness and the triple glazing on the internal skin of the facade. As well as rainwater capture and solar panels, another important sustainability feature is the use of lake water through heat exchange for the heating and cooling of the building.

This strong commitment to sustainability is also reflected in the construction process. Olympic House is an exemplary paradigm of circular economy in construction: 95% of materials from the former administrative buildings on the site were either reused or recycled.

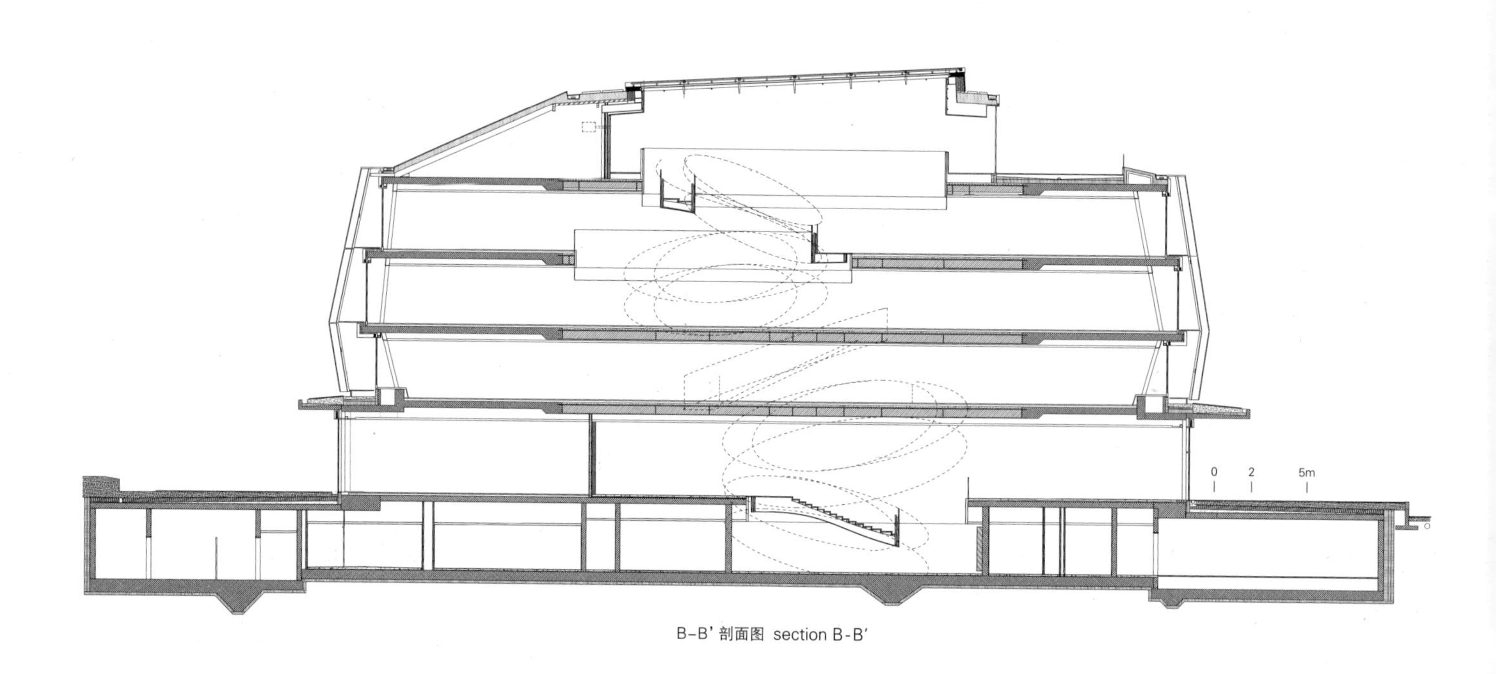

B–B' 剖面图 section B-B'

项目名称：Olympic House-Ioc Headquarters / 地点：Route de Vidy 9 et 11, 1006 Lausanne, Switzerland / 设计建筑师：3XN Architects / 本地建筑师：IttenBrechbühl
项目团队（3XN）：Kim Herforth Nielsen, founder & senior partner; Jan Ammundsen, senior partner in charge of project; Fred Holt, partner; Søren Nersting, project manager
景观设计师：Hüsler & Associés / 室内设计和工作空间规划：RBSGROUP / 建筑照明设计：Jesper Kongshaug / 厨房设计专家：SCHÉMA-TEC / 土木工程师：INGENI SA – Ingenierie Structurale / 暖通空调：Weinmann-Energies SA / 电气工程师：MAB-Ingenierie SA / 立面工程师：Emmer Pfenninger Partner AG / 可持续工程师：ThemaVerde
声效专家：d'Silence acoustique / 防火专家：Ignis Salutem SA / 客户：The International Olympic Committee / 项目总监：Marie Sallois / 建筑设备经理：Thierry Tribolet
项目经理：Nicolas Rogemond / 用地面积：31,390m² / 建筑面积：Olympic House–5,950m²; Château–325m² / 总建筑面积：22,000m² / 总体积：Olympic House–135,000m³; Château–3,900m³ / 建筑规模：Olympic House–6 floors; Château–3 floors / 工作站：500 / 停车场：203 for cars, 25 motorcycles / 施工开始时间：2016.5
竣工时间：2019.6 / 摄影师：©Adam Mørk (courtesy of the architect)

阿迪达斯“体育世界”办公大楼
Adidas World of Sports ARENA

Behnisch Architekten

阿迪达斯“体育世界”办公大楼为 2000 名员工提供灵活可持续的办公空间

New Adidas ARENA Headquarters provides a flexible and sustainable office space for 2,000 employees

阿迪达斯体育世界园区的新办公楼和接待楼，以其强烈的视觉特征吸引着进入园区的游客。该建筑既实用又富有表现力，其雕塑一般的造型强调了阿迪达斯所代表的对运动的热情。大楼占地面积约52000m²，在灵活性和有机性原则的指导下，为近2000名员工提供了一个现代化、可持续的工作空间。

体育世界园区的建设始于1999年，当时这里是一个美军基地。园区包括北部和南部两个部分；第一座新建筑是“条纹”餐厅，然后是“蕾丝”办公大楼、一个健身房和两个多层停车场。随着“半场”餐厅、会议中心和南部园区总部办公大楼的建成，体育世界园区的扩建工作已经完成。

T形的绿色区域是娱乐区，一条延展的交通循环路线将园区的北部和南部连接起来，帮助员工和游客确定方位。

总部办公大楼这座建筑在透明度、景观性和现代感之间取得了完美的平衡，这种现代感来源于跨专业领域的沟通与互动，以及对未来发展的响应。

其成果是惊人的：在模拟景观上，矗立着一座富有抽象性质的建筑，该建筑内含有三层的工作区域。这座雕塑般的山丘容纳了建筑的入口区域，其中一部分作为公共入口，能通向宽敞明亮、适用于各种活动的中庭。

新建筑中使用的材料是出于清晰度和透明度的考量，表现出一种直白性。简单而粗野的细节（例如，管道外露）所营造的工作氛围强调了适应性这一理念。

在中庭里有一段引人注目的悬浮楼梯，通过宽敞明亮的天井一路通向三层的工作区。中央的“主干道”与三层楼纵横相连，打造出一个类似于悬在半空中、热闹的集市感空间。

在每层楼上，“主干道”由六个厨房中控枢纽相连，代表了六个“主要区域”，这些中控枢纽通过不同的材料、颜色和家具类型加以区分，包括柜台和吧台、墙壁构件、吊顶、储物柜、固定式和独立式家具以及照明设备。洛杉矶、伦敦、东京、纽约、上海和巴黎等这些大都市，是阿迪达斯品牌的重要选址地。先是形成社区，再延伸成工作区，激发了本地认同感的发展。在这座大楼里，工作区、正式和非正式会议区、娱乐区相互交替分布，这些区域就像是斑驳的簇团，大小不一，环绕在天井周围，以寻求更多的自然光照。

大楼立面配备了专门研发的光控和遮阳系统，能适应立面的各个朝向，优化了最大透光率和最小热增益的相互作用。固定的外遮阳系统主要由安装在铝框架上的不透明穿孔板组成。由于使用了最新的能源效率标准，并广泛使用了可回收材料，该建筑已提交获取LEED黄金认证。

建筑设计采用了钢筋混凝土结构和倾斜钢筋混凝土支撑系统，这对结构工程来说是一个重大挑战。

The new office and reception building for Adidas' World of Sports campus welcomes visitors to a headquarters with a strong visual identity. The Arena's architecture is both functional and expressive, its sculptural shape emphasizing the passion for sports that Adidas represents. A floor area of 52,000m² provides nearly 2,000 employees with a modern and sustainable workspace organized on flexible, organic principles.

Development of the campus began in 1999, on the site of a former US army base. It consists of a northern and a southern section; the first new building was the Stripes restaurant, followed by the Laces office building, a gym, and two multi-story car parks. With the addition of the new Halftime restaurant and conference center, and the Arena headquarters on the southern campus, the expansion of the World of Sports is now complete.

A T-shaped green area provides a recreational zone, while an extensive network of circulation routes links the northern and southern parts of the campus, helping employees and visitors to navigate around.

The architecture of the Arena building is a finely balanced interplay between transparency, landscape, and a modern concept of work based on communication and interaction across professional fields as well as responsiveness to future developments.

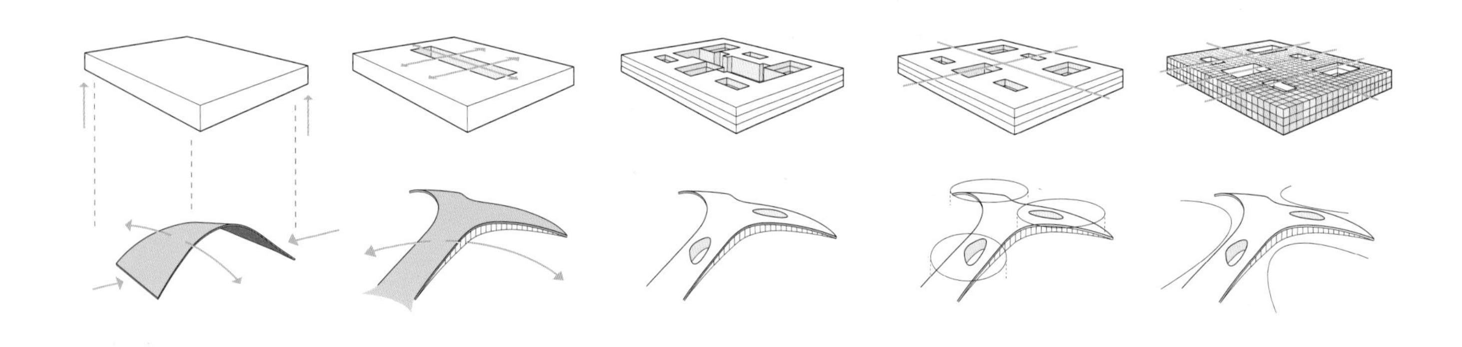

campus north
Green T
Arena
Halftime
main entrance
campus south
N
0
20
50m

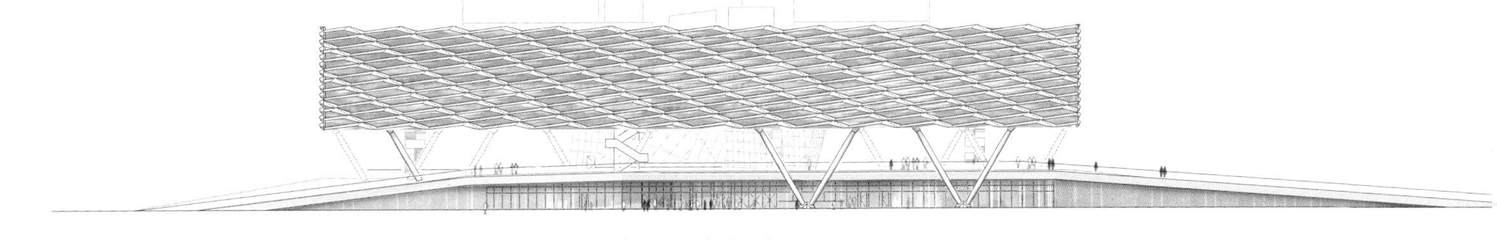

南立面 south elevation

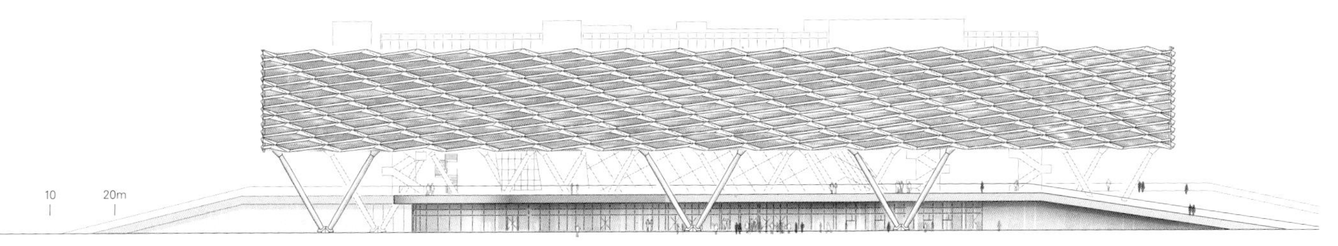

西立面 west elevation

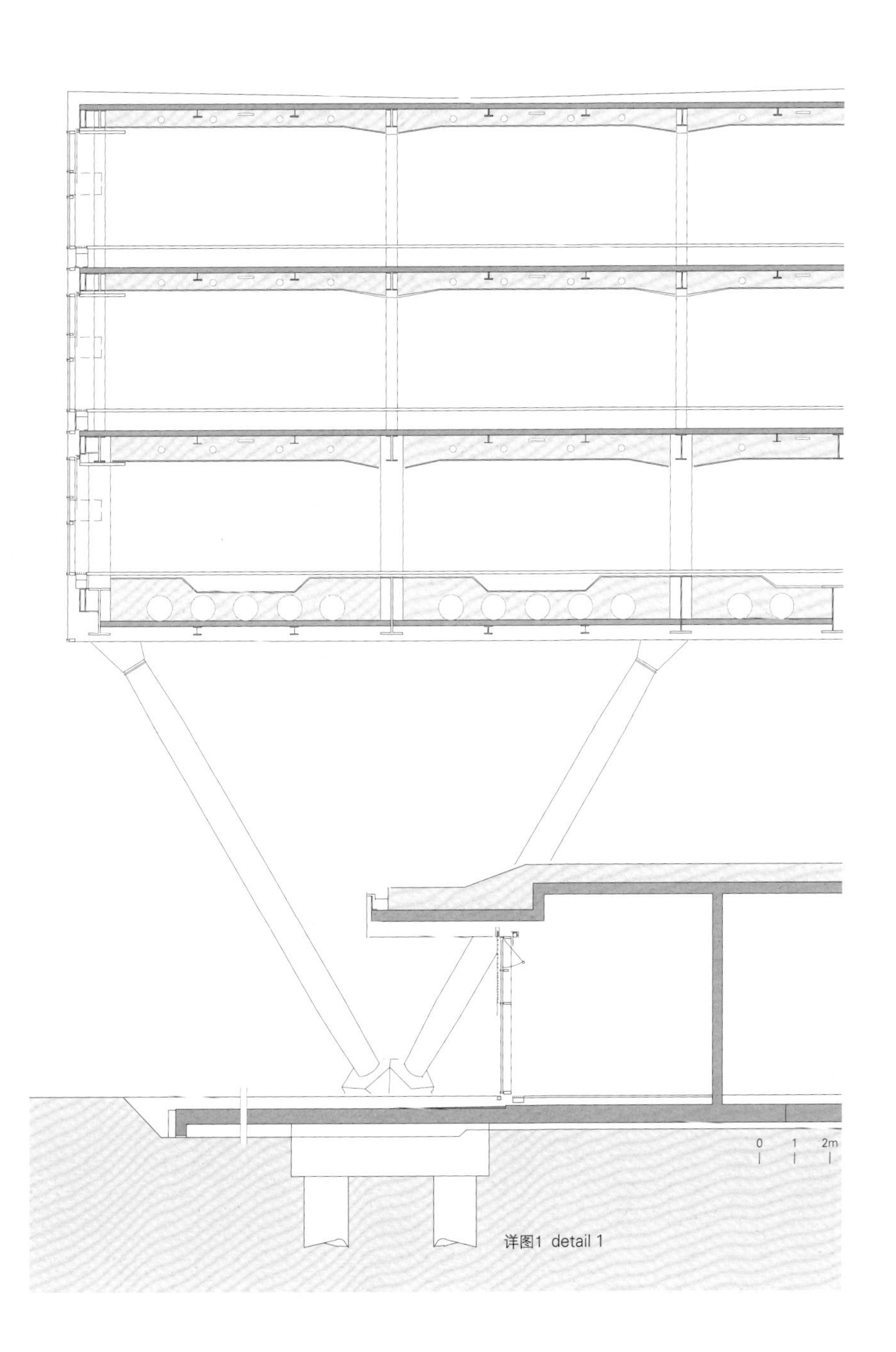

详图1 detail 1

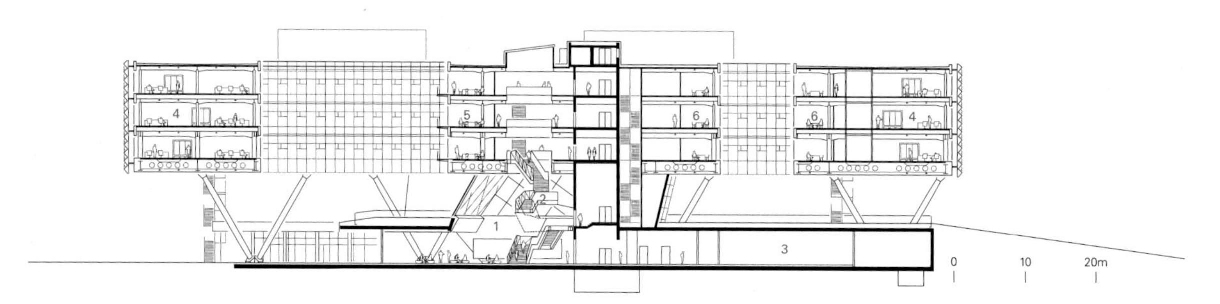

A–A' 剖面图 section A-A'

1. 中庭、半公井区 2. 雕塑楼梯 3. 场馆工作区 4. 开放办公空间 5. 主要街道的特殊场所：游戏室、放映室和图书室
6. 会议室 7. 小酒馆 8. 入口车道 9. 公共区域和自助登记柜台 10. 二三层之间的露天看台 11. 厨房中控枢纽

1. atrium, semi-public zone 2. sculptural staircase 3. back of house 4. open workspace
5. special places on main street: game room, project room, and library 6. meeting rooms 7. bistro
8. access running track 9. public zone and self check-in desk 10. bleachers from 2nd to 3rd floor 11. kitchen hub

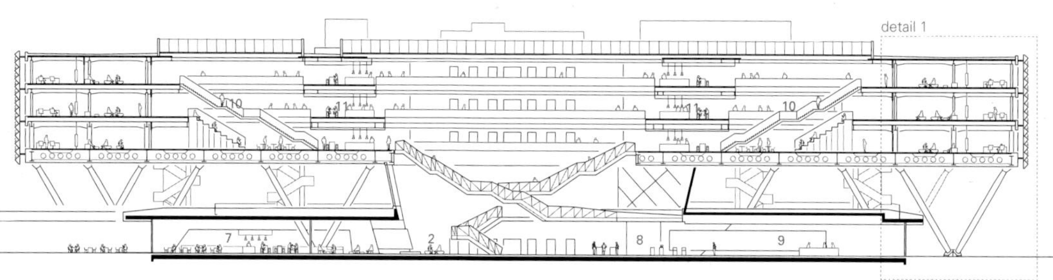

B–B' 剖面图 section B-B'

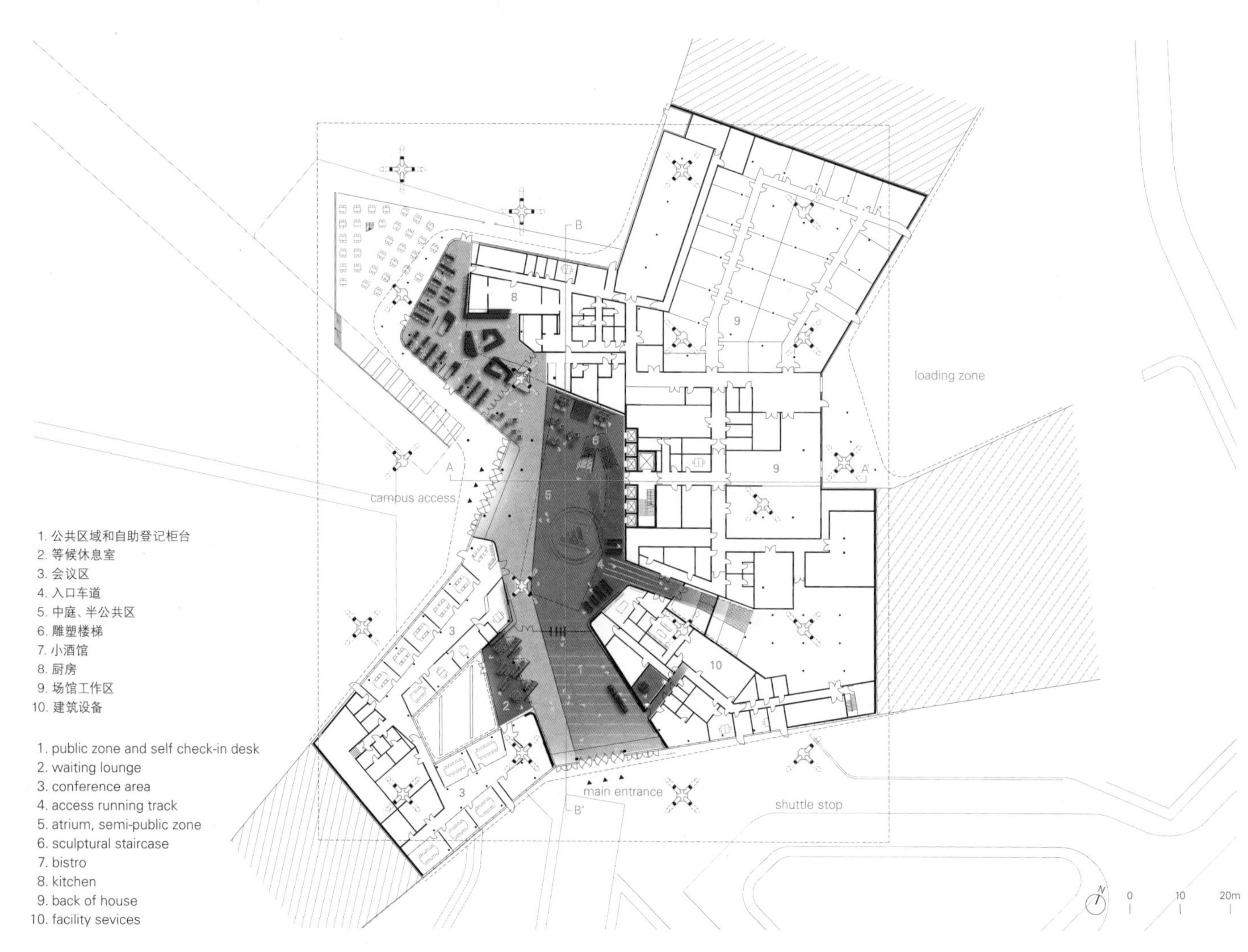

1. 公共区域和自助登记柜台
2. 等候休息室
3. 会议区
4. 入口车道
5. 中庭、半公共区
6. 雕塑楼梯
7. 小酒馆
8. 厨房
9. 场馆工作区
10. 建筑设备

1. public zone and self check-in desk
2. waiting lounge
3. conference area
4. access running track
5. atrium, semi-public zone
6. sculptural staircase
7. bistro
8. kitchen
9. back of house
10. facility sevices

一层 first floor

1. 主要街道
2. 二三层之间的露天看台
3. 主要街道的特殊场所：
   游戏室、放映室和图书室
4. 厨房中控枢纽
5. 衣帽间
6. ID扫码入口
7. 盒子式的电话、聚会、聊天室
8. 会议室
9. 开放办公区

1. main street
2. bleachers from 2nd to 3rd floor
3. special places on main street:
   game-, project- rooms, and library
4. kitchen hub
5. lockers
6. ID-portal
7. phone-, focus-, chat- boxes
8. meeting rooms
9. open workspace

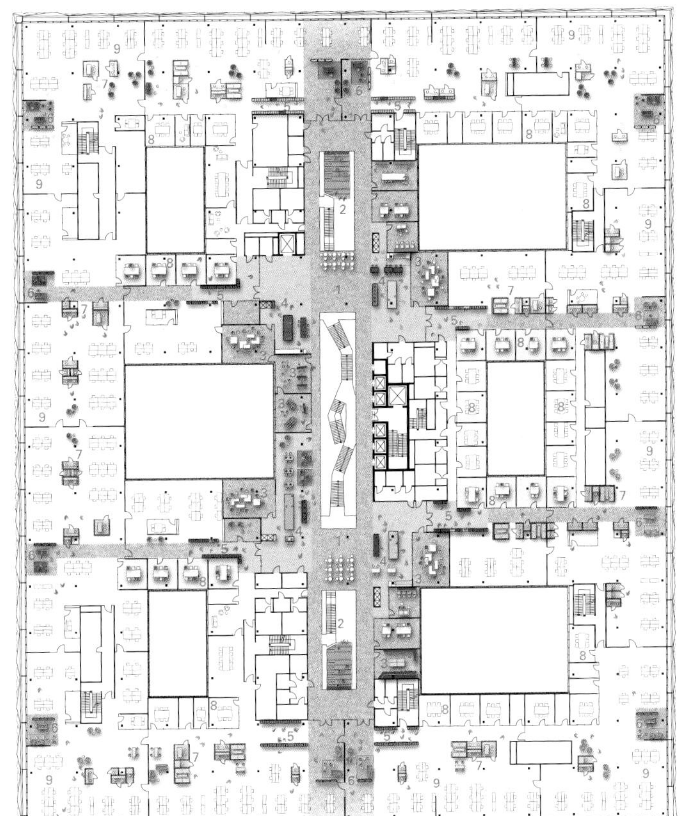

三层 third floor

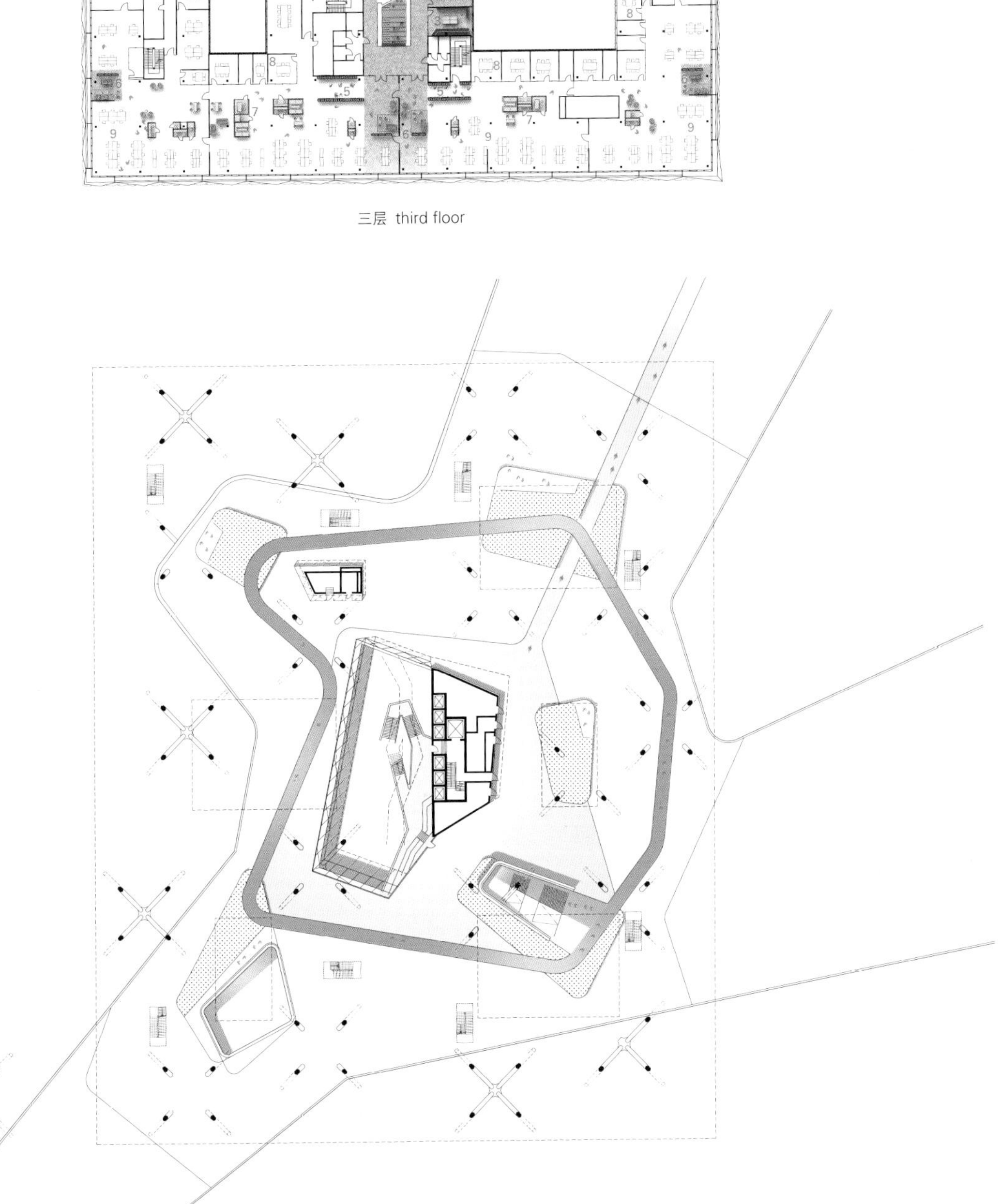

景观层 landscape floor

The result is striking: an abstract volume housing three workspace floors hovers above a modeled landscape. This sculpted hill accommodates the building's entrance area, part of which is publicly accessible, leading to a vast bright atrium suitable for a variety of events.

The materials used in the new building were chosen for legibility, clarity, and expressive straightforwardness. The workshop atmosphere engendered by simple and robust details such as exposed pipework reinforces the idea of adaptability.

A dramatic floating staircase sits in the atrium, rising through the bright and spacious light well to the three-story workspace section. A central "main street" links these three levels both vertically and horizontally, creating a space akin to a lively marketplace suspended in mid-air.

On each floor, the "main street" is adjoined by six kitchen hubs representing six "key cities" differentiated by distinctive materials, colors, and types of furnishing, including counters and bars, wall elements, dropped ceilings, lockers, fitted and freestanding furniture, and lighting fixtures. The selected metropolises – Los Angeles, London, Tokyo, New York, Shanghai, and Paris – are important locations for the Adidas brand. Neighborhoods take shape and extend into the workspace, encouraging the development of local identities. On all three floors, workspaces alternate with areas for formal and informal meetings and recreational zones. They are organized as variegated clusters around lightwells of different sizes, optimized to supply ample natural light.

The facade was equipped with a specially developed sun control and shading system, adapted to the orientation of each side, which optimizes the interplay between maximum light transmittance and minimum heat gain. The fixed external shading system mainly consists of opaque and perforated sheets mounted on an aluminum frame. Thanks to its implementation of up-to-date energy-efficiency standards and extensive use of recyclable materials, the building has been submitted for LEED Gold certification.

The architectural design, which represented a significant challenge in terms of structural engineering, was realized using steel and concrete construction and a system of slanted steel and concrete supports.

adidas
ESTD. 1949
HERZOGENAURACH
THROUGH SPORT WE HAVE THE POWER TO CHANGE LIVES

项目名称：Adidas World of Sports ARENA / 地点：Adi-Dassler-Straße 1, 91074 Herzogenaurach, Germany / 建筑师：Behnisch Architekten / 合作伙伴：Stefan Rappold / 项目主管：Cornelia Wust / 项目团队：Nadine Hoss, Carina Steidele, Dennis Wirth, Nevyana Tomeva, Martin Buchall, Jorge Carvajal, Laetitia Pierlot, Saori Yamane, Adriana Potlog, Ioana Fagarasan, Anna-Lena Wörn, Abdalrahman Alshorafa, Arlette Haker, Matteo Cavalli, Mahboubeh Shoeybi, Hamdy Saflo, Andreas Peyker, Nadine Waldmann / 结构工程：Werner Sobek Stuttgart GmbH / 机电暖通：Rentschler und Riedesser
电力和人工照明：Raible + Partner GmbH & Co. KG / 消防：Endreß Ingenieurgesellschaft mbH, Oehmke + Herbert / 建筑物理：Bobran Ingenieure
立面：KuB Fassadentechnik / 特殊区域的照明设计：Bartenbach GmbH / 园区景观：LOLA Landscape Architects / 总承包商：Ed. Züblin AG / 客户：adidas AG
总建筑面积：52,000m² / 体积：259,000m³ / 竞赛、设计开始时间：2014 / 施工开始时间：2016 / 竣工时间：2019 / 摄影师：©David Matthiessen (courtesy of the architect)

30
PHONE ROOM

2NYC 625
2NYC 628
2NYC 629
2NYC 626
2NYC 569
2NYC 572
2NYC 573
2NYC 570

55
PHONE
ROOM

加兰蒂 BBVA 银行科技园区

Garanti BBVA Bank Technology Campus

ERA Architects

METRO

坐落在伊斯坦布尔一座老工业厂房内的新 BBVA 环保办公园区

BBVA's new eco-friendly office campus sits in an old industrial factory complex in Istanbul

该项目位于两条主要公路之间，靠近伊斯坦布尔彭迪克机场，由一家老旧的工业化工厂改造成了现在的土耳其银行技术园区。

建筑师的灵感来自于场地周围的自然地形，建筑师希望能与周围的城市结构形成对比。建筑体的形状如同水晶，平缓地坐落在几个人工山丘上。它包含多种功能，如两个礼堂、教育会议室、自助餐厅、休息室、数据中心等。总建筑面积约142000m²。

水平浮动体量通过悬桥和玻璃幕墙连接两块土地，跨度超过30m，建筑主体结构为办公楼层的混凝土平板结构，和下级楼层的裸露框架结构。沿着建筑外表面，深梁悬臂释放了更多的办公空间。所有办公楼层的结构都很相似；置于屋顶的大型组合梁支撑了最宽的悬臂区域。第四层办公区域位于转换结构上，从而构成了位于山丘之中的大礼堂。

灵活性和日光最大化是工作空间设计的主要标准；通过一系列抬高的开放式庭院来使工作区域更具开放性，在这些庭院里，员工可以到处走动，还可以在桥梁上的休息室中休息和交谈，也能欣赏不同的景观。这种透明性将工作空间与城市结合起来。

架空的水平透明体量呈现锐利干练的悬挑形象，使得天空、景观和城市能反映在建筑的立面上。然而，内部主轴结构却是围绕挑空空间建造的，与外部形成了对比。

在人工山丘之间，有一条像小溪一样起伏的小路，用户可以沿着这条小路行走于室外。抬高的体量使得人们能继续这段旅程般的体验。景观内的池塘增强了对周围环境的冷却效果，还有玻璃立面，也可以减少在炎热季节中温度的上升。

该建筑贯彻可持续发展原则，比如，带有动态遮阳和照明的单双层外墙系统，以及为内部空间提供新鲜空气的特殊供暖和冷却系统。考虑到了大楼外交通繁忙的高速公路，外墙被设计成了一个统一的系统，实现了高质量保温与隔声的双重效果。

在最初的设计阶段，场地到手时，拆除了原有的化工厂，检查了土壤是否受污染。剩下的部分被分解、分类，并相应地送去回收。老工厂周围现存的松树被保留了下来，其中一些被移植到新的厂址内。收集的雨水用来灌溉周围的景观。一系列嵌入景天属植物表面的电缆系统形成了绿色的山体。这是土耳其最大的绿色屋顶系统应用之一。因此，此园区获得了LEED黄金认证。

Located between two major highways and near the airport in Pendik, Istanbul, the project transforms an old industrial chemical factory into a technology campus for a dynamic Turkish bank.

The architects were inspired by the natural topography around the site and the desire to create a contrast with the surrounding urban fabric. A crystal volume, the main working environment, lies gently over several artificial hills. This shelters various functions such as two auditoriums, educational meeting spaces, cafeterias, lounges, a data center and more. The total built area is approximately 142,000m².

The horizontal floating volume connects two plots of the site by the introduction of hanging bridges and glazed screen walls, spanning over 30m. The buildings' main structure is a flat concrete slab system at the office level, and an exposed frame structure at the lower levels. Along the outer skin, office spaces have been liberated by deep cantilevers. All four office levels are similar; the widest cantilevering areas are supported by large composite beams placed on the roof. The four-story office block stands partially on a transfer structure to enable

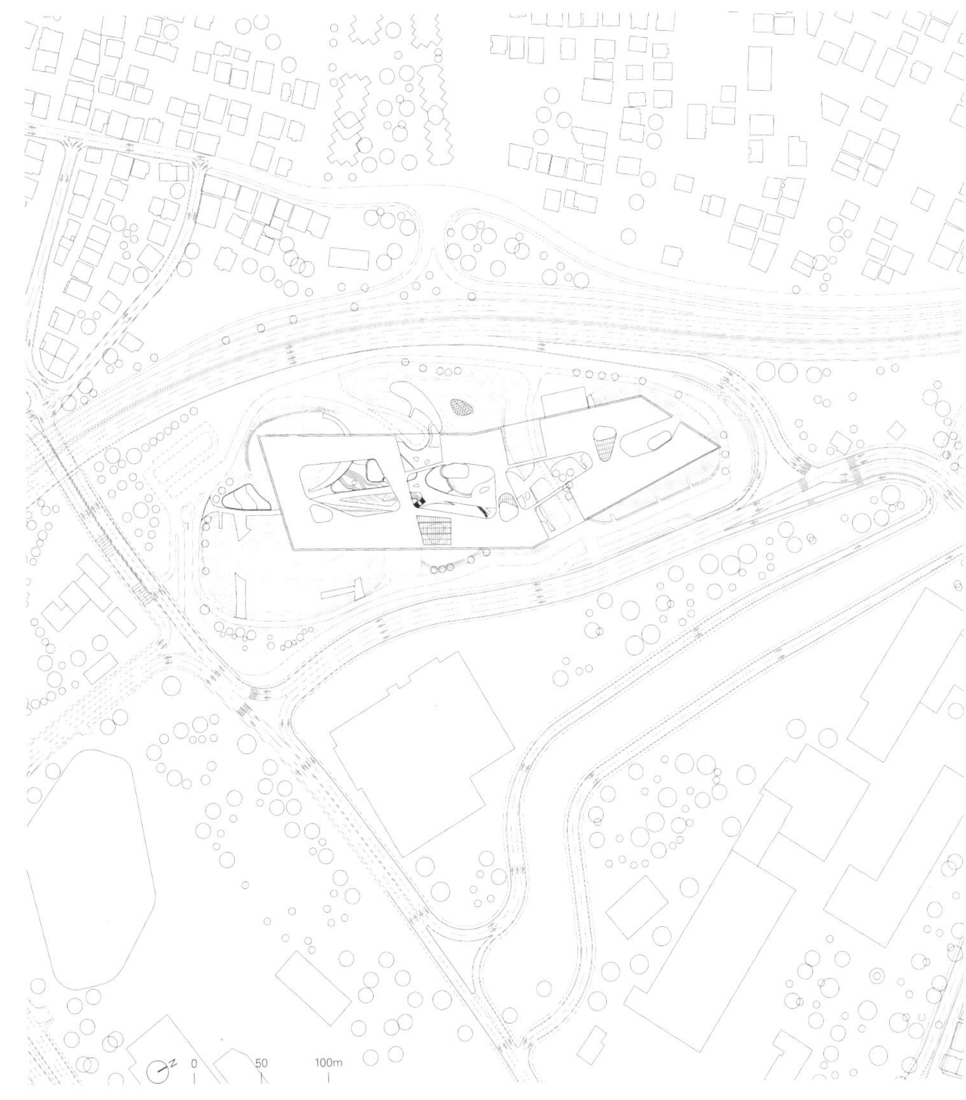

项目名称：Garanti BBVA Bank Technology Campus
地点：Pendik, Istanbul, Turkey
建筑师：ERA Architects – Ali Hızıroğlu
建筑团队：Ertun Hızıroğlu, Çiğdem Duman, Faik Barutçu, Fahrettin Çiloğlu, Sevgi Paçcı, Serap Aydınel, Adnan Günay, Gökçen Demirkır, Melis Uysal, Çağla Özgen, Derya Elhan, Ekim Orhan Ismi, Süreyya Numanoğlu, Elif Özil Gülizar Bozdoğan, Zuhal Özen, Şafak Kızılırmak
室内设计：Midek Mingü; ERA – Hasan Mingü
结构设计：Mpi, Melih Bulgur
电气设计：Proma, Erdoğan Duman
声学顾问：Talayman Consultants
立面、结构、机电暖通、可持续性、LEED标准顾问：Werner Sobek
立面设计顾问：Tim Macfarlane, Glass Limited
照明设计：Planlux, Speirs and Major
客户：Garanti Bank BBVA
用地面积：53,000m²
总建筑面积：142,000m²
设计时间：2010
竣工时间：2018.9
摄影师：©Cemal Emden (courtesy of the architect)

the wide span volume for the auditorium which is located within one of the hills.
Flexibility and maximizing daylight were major criteria for the working spaces; accessibility to the office spaces has been organized through a set of elevated open courtyards where the users circulate and also have the opportunity to rest and socialize in lounges placed on bridges, which allow for many surprising vistas. This transparency is used to integrate the working spaces with the city.
A strong contrast awaits the visitor, as the elevated horizontal crystal has a sharp and decisive cantilevered appearance from the outside, reflecting the sky, the landscape and the city. Yet the inner main axis is, conversely, structured around voids. The user circulates through the outdoor ground levels along a path which undulates like a creek between the artificial hills. The elevated volume has been designed to continue this experience of a journey. Ponds within the landscape enhance the cooling effect on the environs, including the glazed facades, to reduce temperature gain during hot seasons.

The building implements sustainable principles, from single and double skin facade systems with dynamic sun shading integrated with lighting, to special heating and cooling systems providing the inner spaces with fresh air. The façades have been designed as unitized systems, allowing high quality thermal and acoustic insulation, important considering that the site is surrounded by highways with heavy traffic.
During the initial design stages, when the site was acquired, the old chemical factory was removed and the soil checked for contamination. The remains were dismantled, categorized, and sent for recycling accordingly. The existing pine trees surrounding the old factory were kept, with a few transplanted into new locations within the site. Rain water harvesting is used to irrigate the landscaped surroundings. The green hills were achieved through a system of cables embedded into the sedum surface. This is one of the largest green roof system applications in Turkey. As a result of these endeavors he campus has received LEED Gold certification.

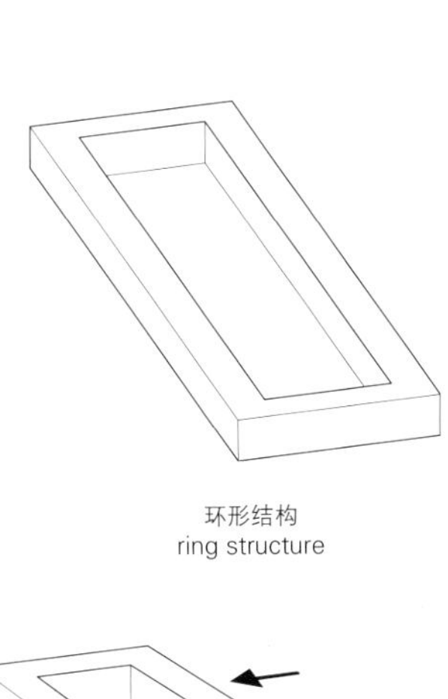

环形结构
ring structure

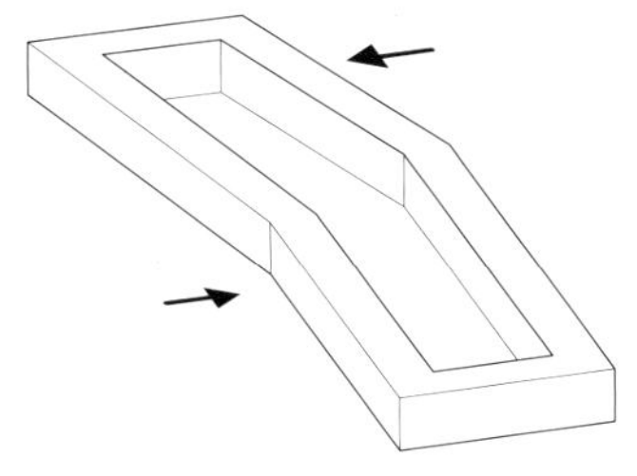

场地适应性
site adaptation

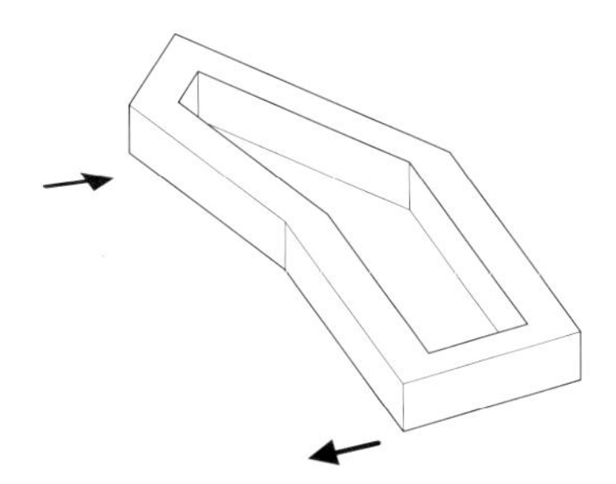

变形
deformation

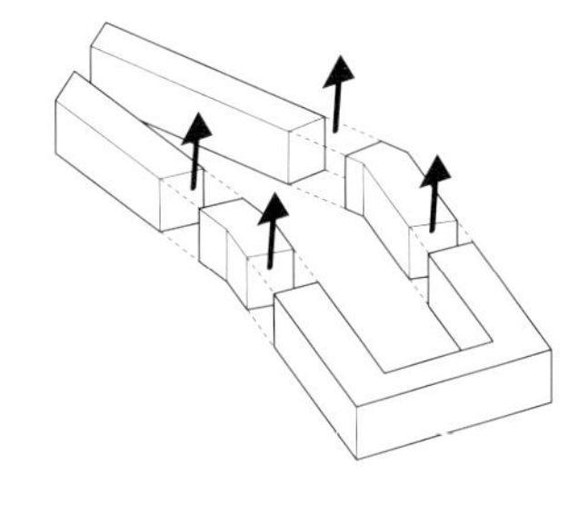

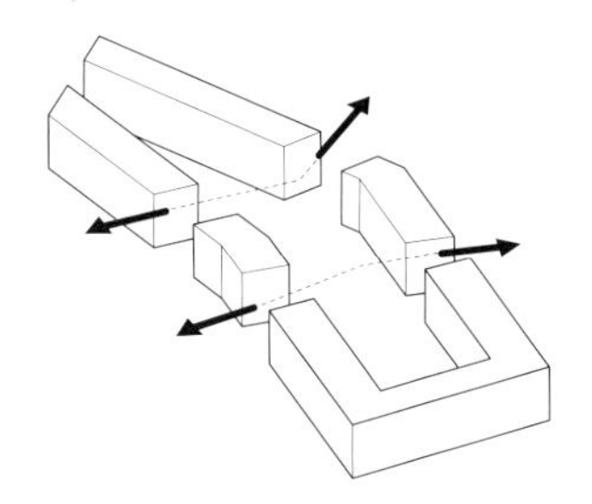

减少
reduction

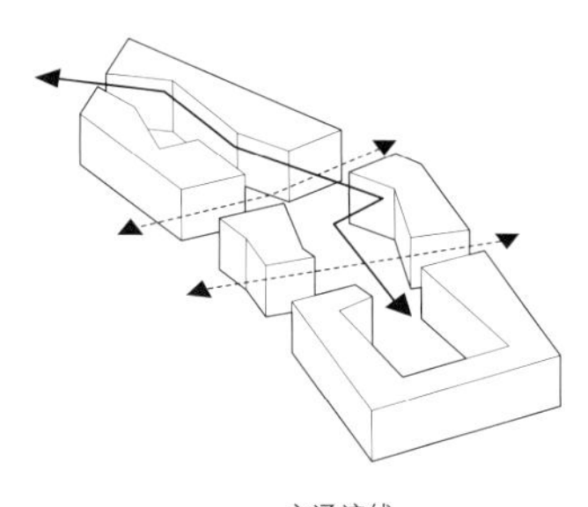

交通流线
circulation

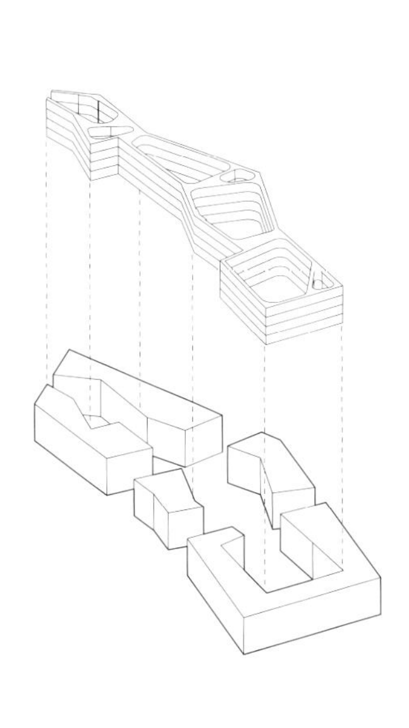

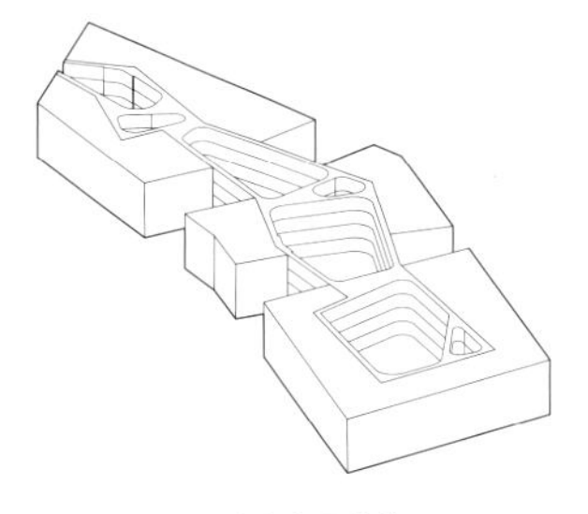

内部交通流线
inner circulation

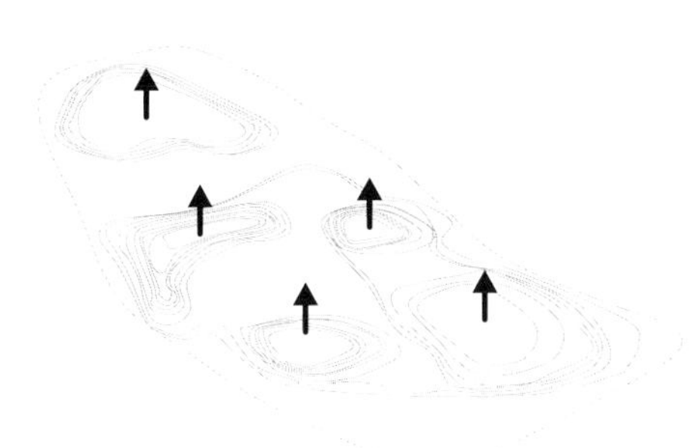

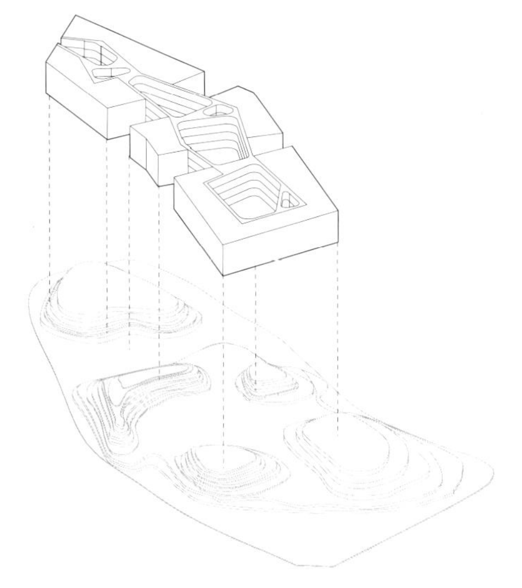

景观设计策略
landscape strategy

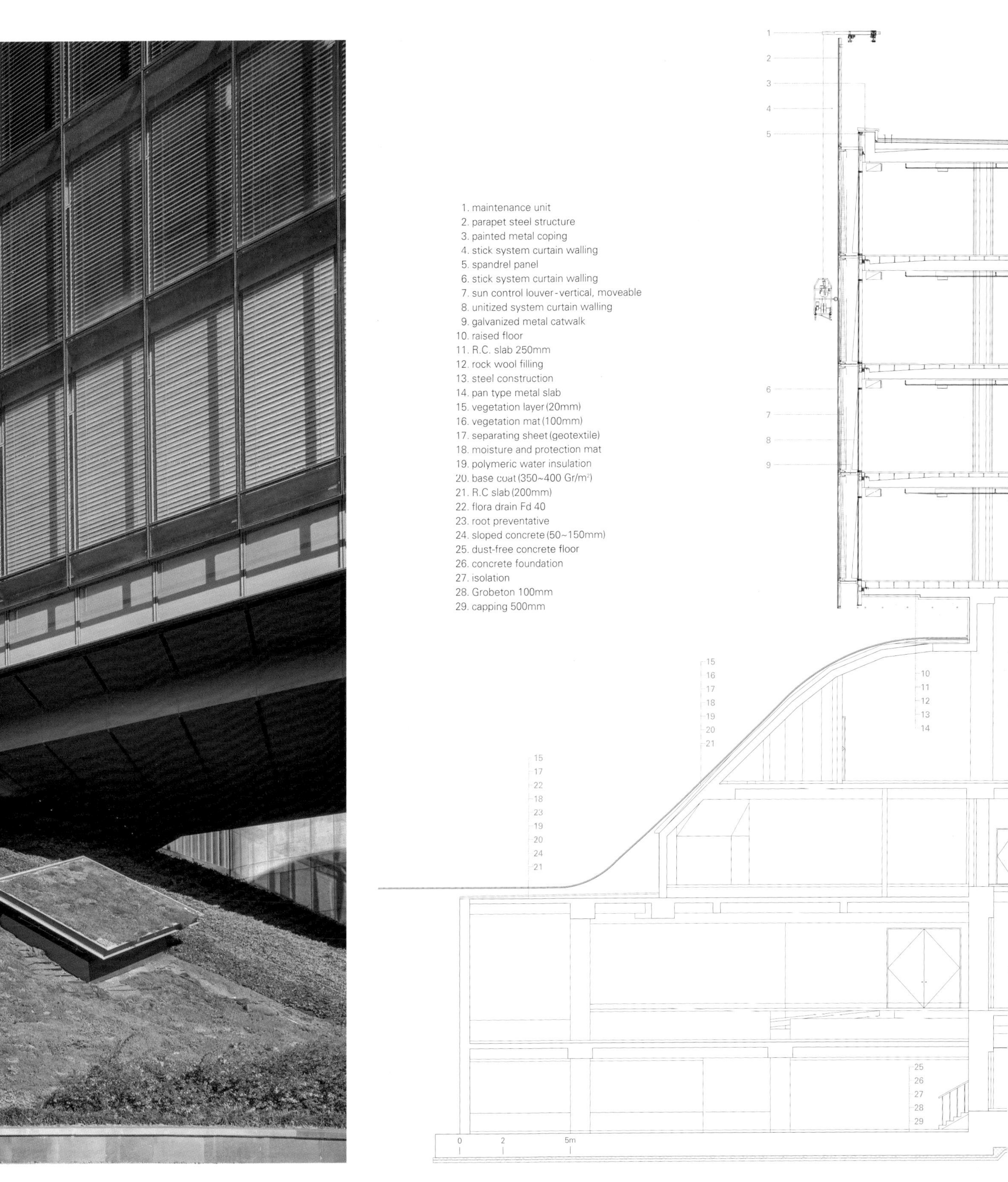
1. maintenance unit
2. parapet steel structure
3. painted metal coping
4. stick system curtain walling
5. spandrel panel
6. stick system curtain walling
7. sun control louver-vertical, moveable
8. unitized system curtain walling
9. galvanized metal catwalk
10. raised floor
11. R.C. slab 250mm
12. rock wool filling
13. steel construction
14. pan type metal slab
15. vegetation layer (20mm)
16. vegetation mat (100mm)
17. separating sheet (geotextile)
18. moisture and protection mat
19. polymeric water insulation
20. base coat (350~400 Gr/m²)
21. R.C slab (200mm)
22. flora drain Fd 40
23. root preventative
24. sloped concrete (50~150mm)
25. dust-free concrete floor
26. concrete foundation
27. isolation
28. Grobeton 100mm
29. capping 500mm
0
2
5m

详图1 detail 1

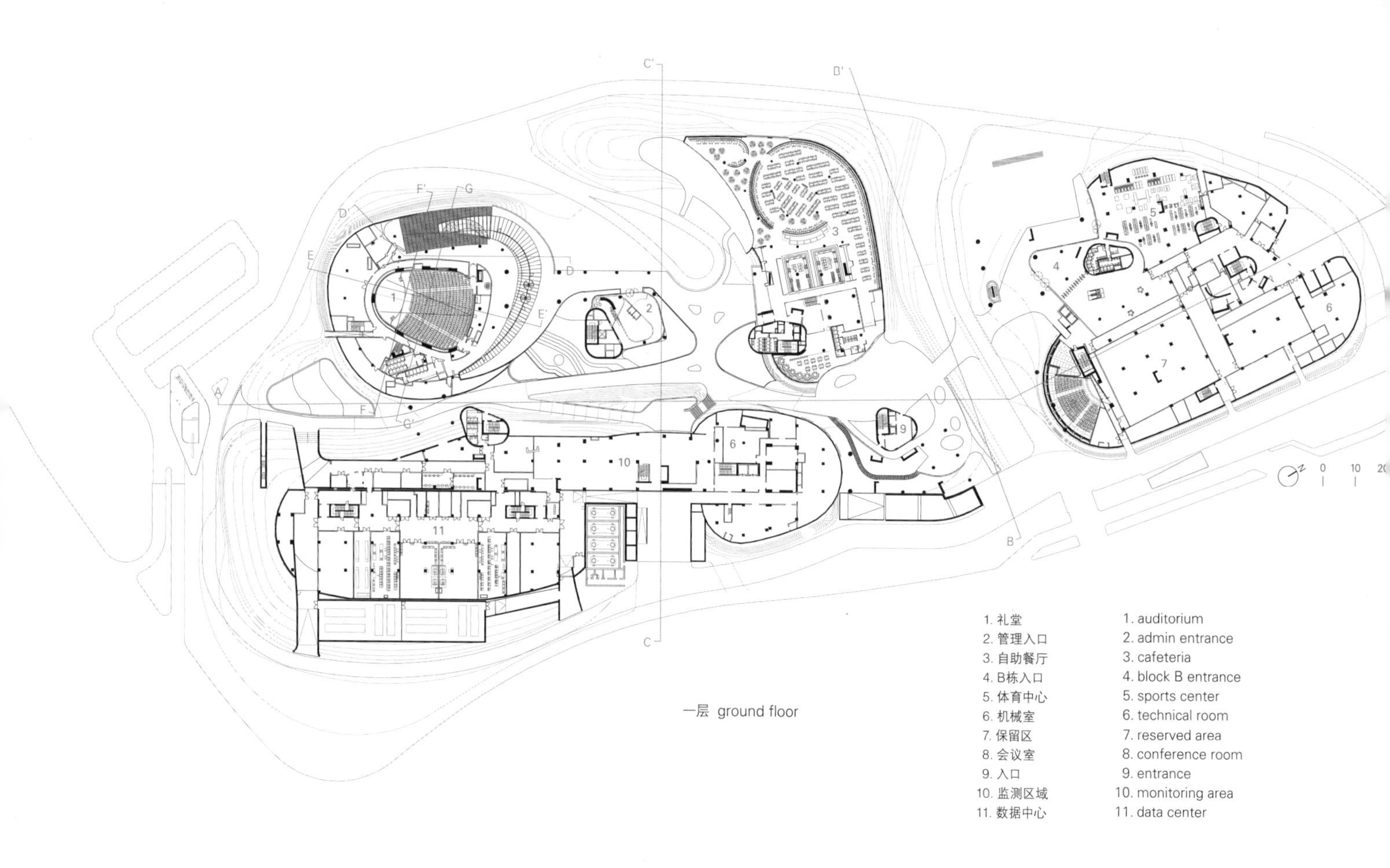

一层 ground floor

1. 礼堂
2. 管理入口
3. 自助餐厅
4. B栋入口
5. 体育中心
6. 机械室
7. 保留区
8. 会议室
9. 入口
10. 监测区域
11. 数据中心

1. auditorium
2. admin entrance
3. cafeteria
4. block B entrance
5. sports center
6. technical room
7. reserved area
8. conference room
9. entrance
10. monitoring area
11. data center

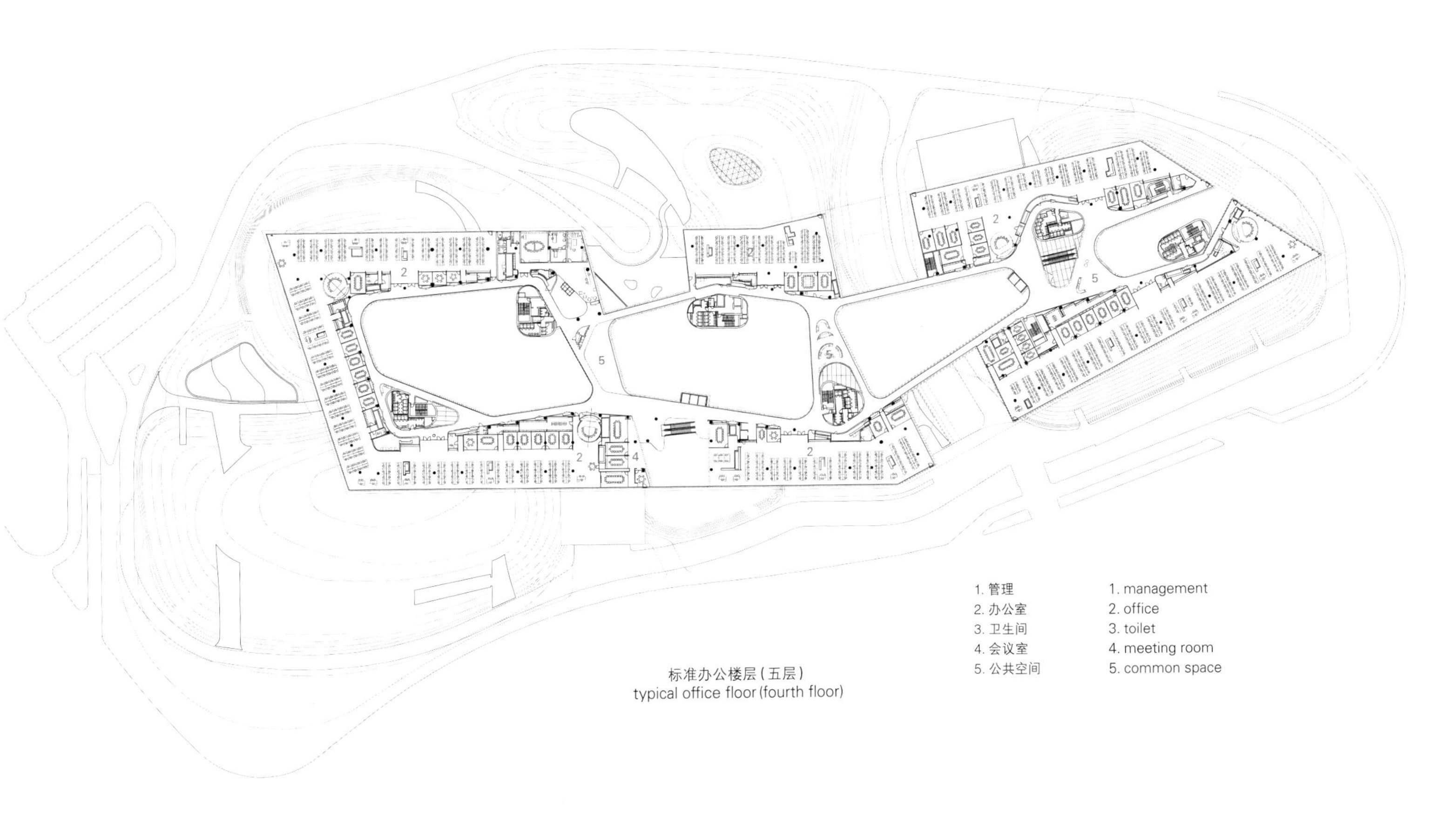

标准办公楼层（五层）
typical office floor (fourth floor)

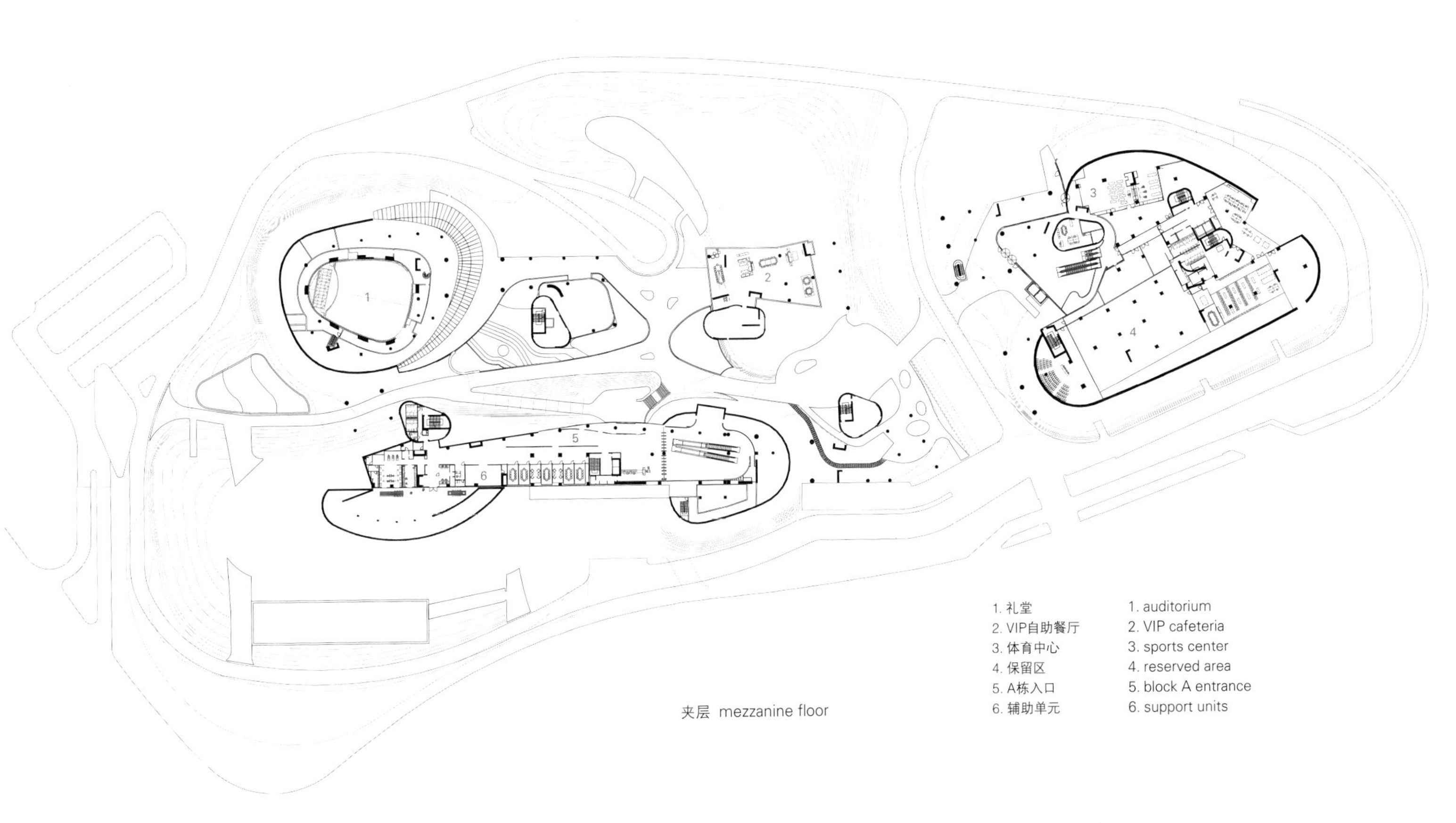

夹层 mezzanine floor

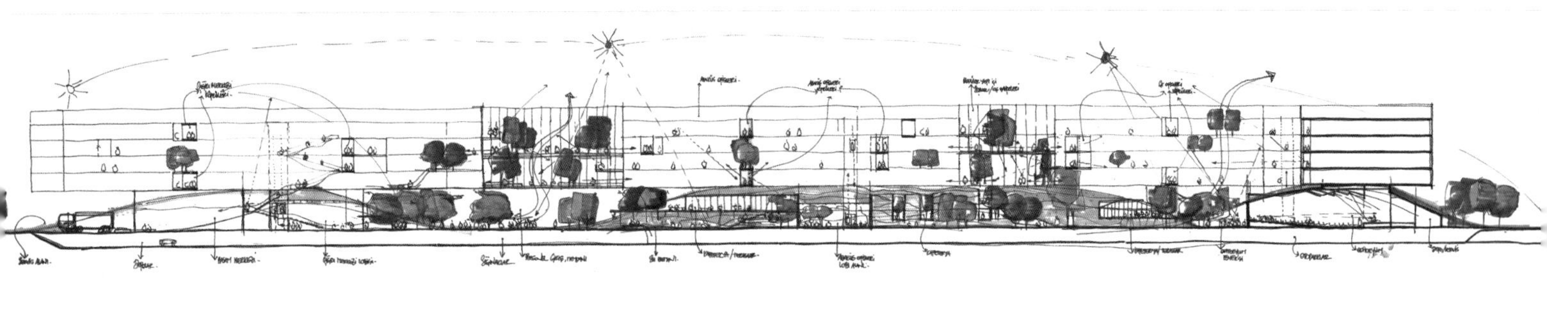

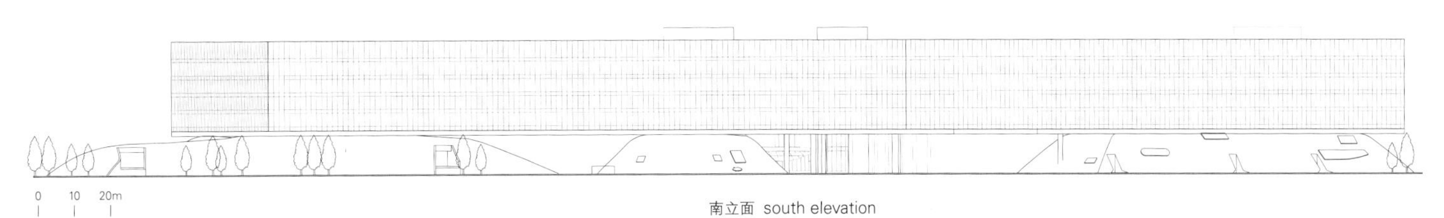

南立面 south elevation

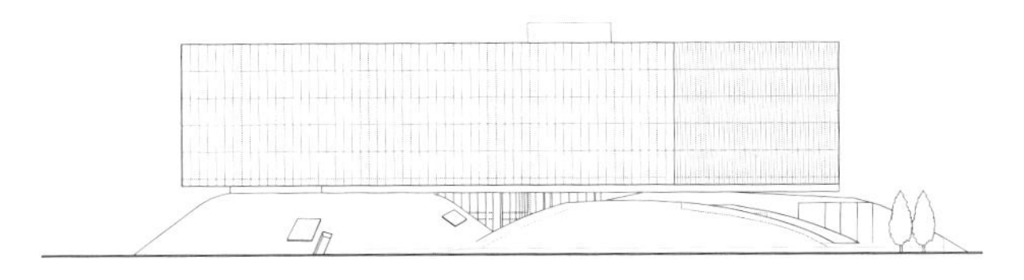

西立面 west elevation

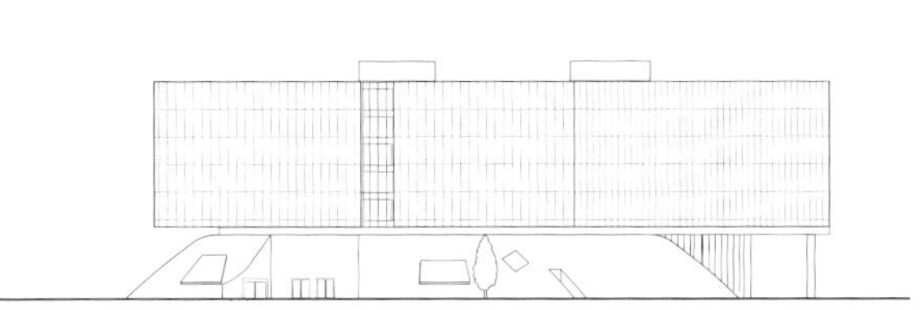

东立面 east elevation

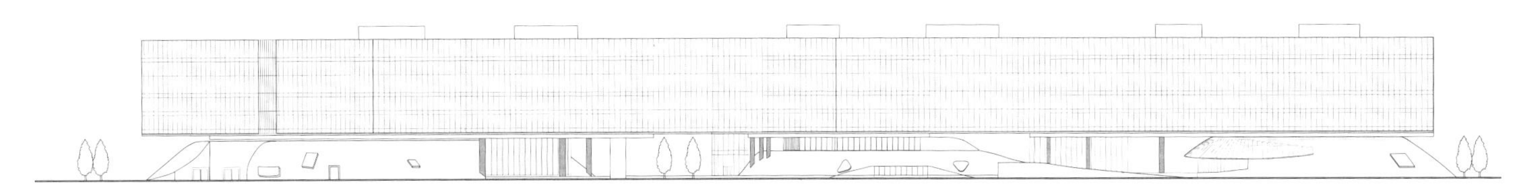

北立面 north elevation

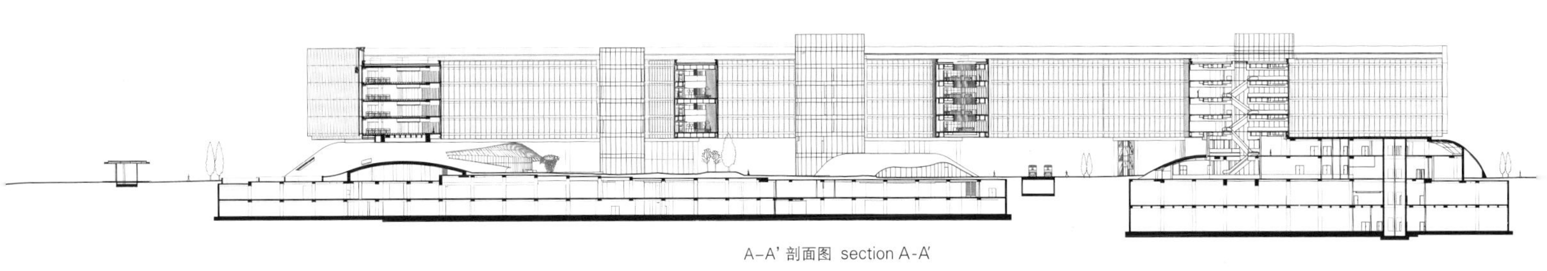

A–A' 剖面图 section A-A'

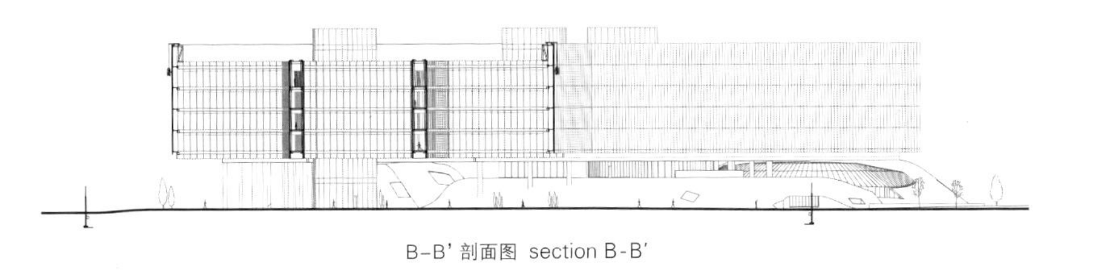

B–B' 剖面图 section B-B'

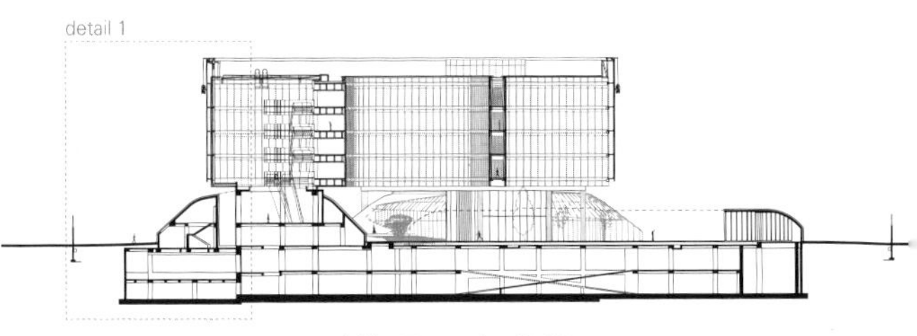

C–C' 剖面图 section C-C'

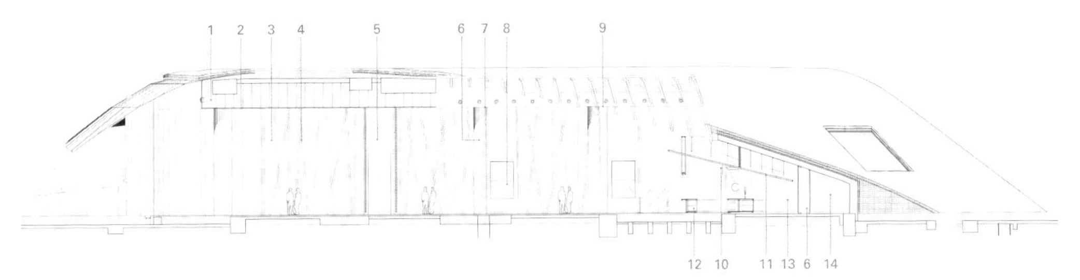

D-D' 剖面图 section D-D'

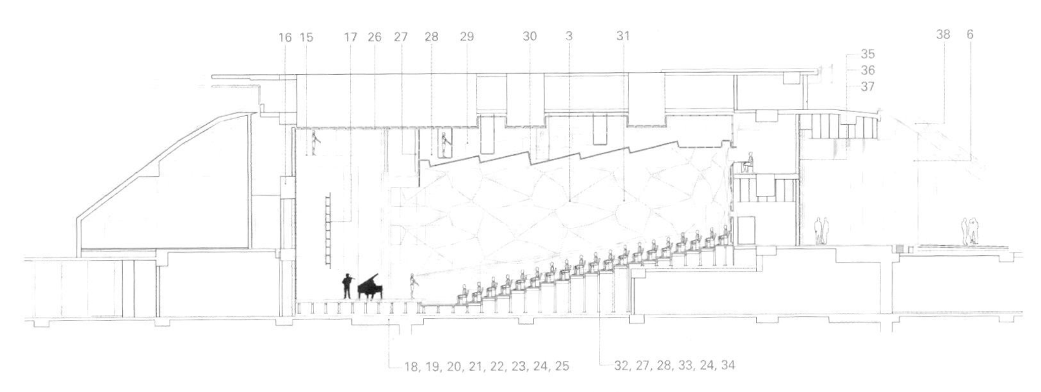

E-E' 剖面图 section E-E'

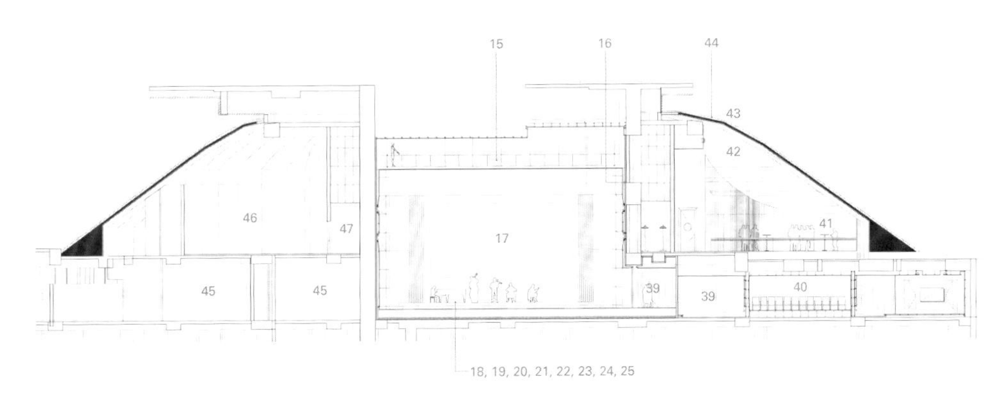

F-F' 剖面图 section F-F'

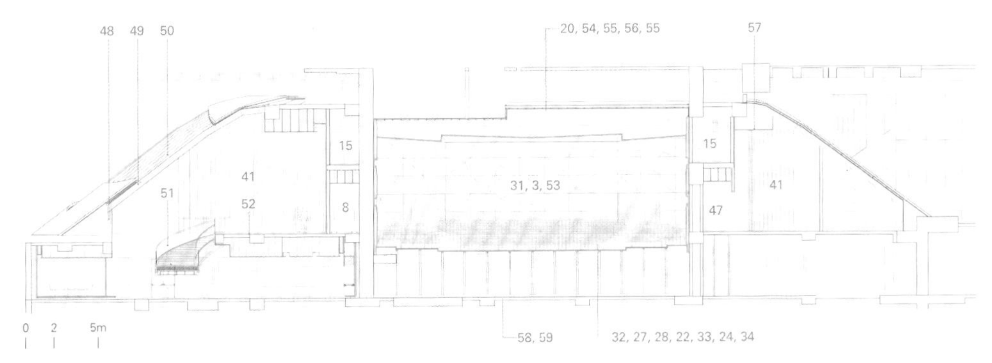

G-G' 剖面图 section G-G'

1. jet nozzle
2. ceiling
3. acoustic panel
4. tube
5. laminate covering
6. concrete colon
7. nozzle
8. cloakroom
9. steel structure
10. lacquer panel
11. ceiling floor
12. bar service
13. wood panel
14. kitchen
15. technical space
16. aqua panel
17. projection screen
18. 20mm wood
19. 24mm contra
20. rock wool filling
21. polyethylene
22. wood timber
23. concrete filling
24. steel carrier system
25. concrete
26. acoustic isolation ceiling
27. acoustic isolation
28. culvert
29. cat walk
30. stage lighting element
31. angled acoustic panel
32. stair floor
33. 12mm contra
34. R.C. slab
35. plaster
36. gas concrete
37. aqua panel paint
38. glass roof
39. foyer basement floor
40. studio
41. foyer
42. wood cover
43. concrete cower
44. green roof
45. mechanic room
46. office
47. corridor
48. concrete structure
49. steel beam
50. glass
51. glass railing
52. natural stone floor
53. lighting element
54. vibration receptors
55. 12.5mm aqua panel
56. isolation panel
57. ceiling acoustic panel
58. 150mm cement finish
59. isolation mattress

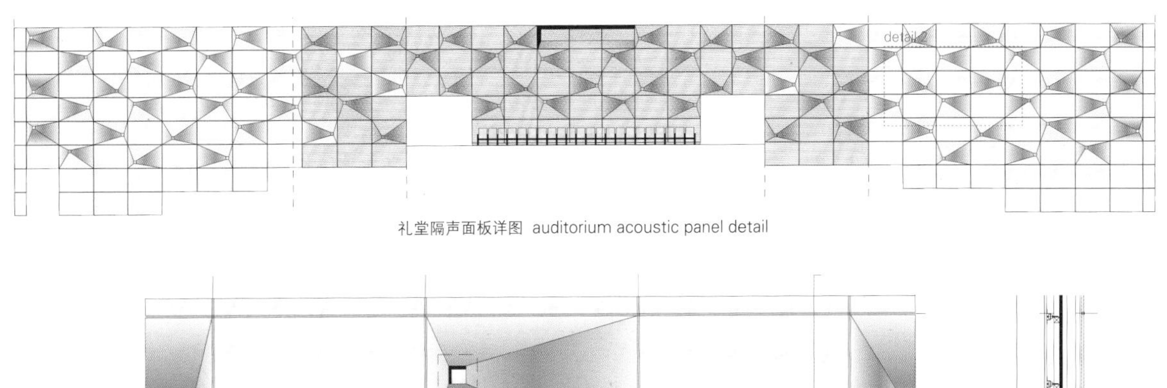

礼堂隔声面板详图 auditorium acoustic panel detail

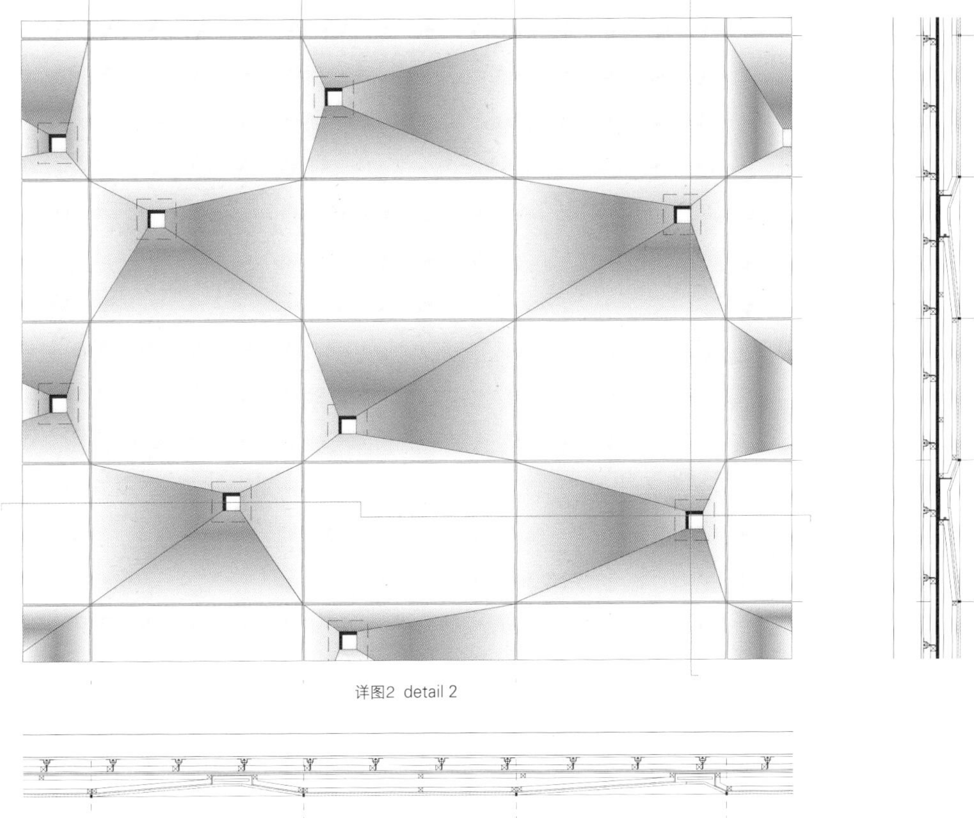

详图2 detail 2

塔塔一体化办公园区
Titan Integrity Campus

Mindspace

塔塔新园区通过与自然的融合，重新诠释了企业的办公理念

Titan's new campus reinterprets the concept of a corporate office by integrating it with nature

2014年，身为生活消费品公司的塔塔公司举行了一场旨在打造新型办公空间的非公开设计竞赛。新办公空间的设计应将建筑环境与自然相结合，在宁静的环境中创造一个令人羡慕、积极健康的新式办公环境。

新办公园区的建筑用地面积为6.5英亩，位于印度班加罗尔市韦拉桑德拉湖畔的电子城中。在园区里另外打造了一个人工的"生态湖"，新的办公楼及其附属功能建筑就围绕着该湖而建。

建筑较长一侧的立面坐南朝北，最大限度地减少了太阳的眩光。建筑材料和造型上的多孔性形成风洞，产生文丘里效应，产生持续的通风，这样整个公共区域都是开放的，不需要安装空调。

这座三层结构在每个楼层都设有露台花园。自由流动的层叠绿色露台通过外部楼梯彼此相连。这些露台还为下方的办公空间起到了保温的效果，从而减少热量/空调负荷，这与可持续建筑的理念产生了共鸣。绿色露台不仅可以让人在户外工作，鼓励员工之间的互动，还可以为那些希望在休息时间悠闲散步的人提供小径。

空间规划——统筹不同部门

项目功能规划是这样的，每个部门，如手表、珠宝、眼镜、配件等都有自己的区域，但这些区域通过中庭相互连接，便于采光，并使热空气逸出。电梯和楼梯都设置在中庭内。办公空间的进深可以保证整个办公空间会布满自然光，在需要人工照明的时候，其使用主要是靠自然光线和感应传感器来实现的，从而最大限度地减少了人造光的使用。太阳能电池板安装在西侧的露台上方和后院的上方，可以现场供能，满足25%的能源需求。

该建筑的一层被微微抬起，瀑布状的水池沿着这一层形成整座建筑的中心轴。

清晨，人们通过水面的倒影可以看到水纹的波动，二者相映成趣，为空间提供了一个额外的维度。一层被抬升了大约2m，使园内的"生态湖"和外面的韦拉桑德拉湖之间无缝衔接，使二者的分界消失。

沿着水体的中心轴有一个线性的双层高空间，与一系列宽阔的台阶、庭院、产品展示墙、座位和日常会议空间相结合。这条轴线的终点就是中庭，从中庭可以通往餐厅，中庭内满是绿色植物。在这里，人们既可以在室内用餐，也可以在室外，聆听鸟鸣和流水的声音。餐厅有三层，可以当作一个多功能大厅使用。

中庭——连接与社区

园区内的五个中庭从底层向上垂直连接所有楼层，引入自然光线并营造出一种社区感。"垂直公园"的景观设计融入了动态的建筑风格，将外部和内部环境无缝衔接到一起。

归根结底，塔塔园区重新诠释了公司办公室的理念，并将其融入大自然中，从而为终端用户提供灵感，并且提升他们的体验。

GREEN TERRACES -OUTDOOR WORKSPACES
ELEVATED GROUND AT EVERY LEVEL
CLEAR WATER BODY / FIRE SUMP
LIGHT WELLS/VISUAL MARKERS / NODES
GREEN WALK/
PATHWAY
BIO LAKE
WATER EDGE
MAIN ACCESS ROAD
MAIN ENTRY
INTERNAL ROAD
OUTSIDE KUCCHA ROAD
VEERASANDRA LAKE

In 2014, consumer goods company Titan held a closed competition for a new corporate office; the design should integrate the built environment with nature to create an enviable, healthy new workplace in a serene setting.
The 6.5 acre site is located in Electronic City, near Veerasandra Lake, Bangalore. A second, adjoining "bio-lake" was created, around which the new office and its ancillary functions are based.
The building is oriented with the longer sides facing north to south, minimizing solar glare. Porosity in materials and form generates a continuous breeze, with wind tunnels creating a Venturi effect, allowing common areas to be open and non-air conditioned.
The three-floor structure has terrace gardens at every level. Free-flowing cascading green terraces are connected by external staircases. These terraces provide insulation to office spaces below, reducing heat and AC load, and making the building more sustainable. The terraces facilitate outdoor working, stimulate interaction between employees and also offer trails for leisurely walks in breaktimes.

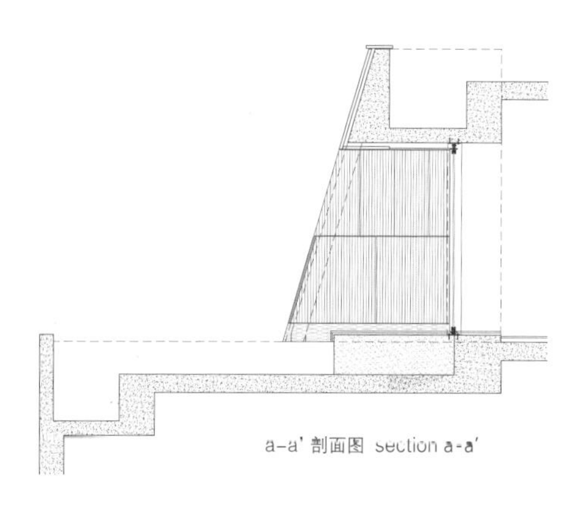
a-a' 剖面图 section a-a'

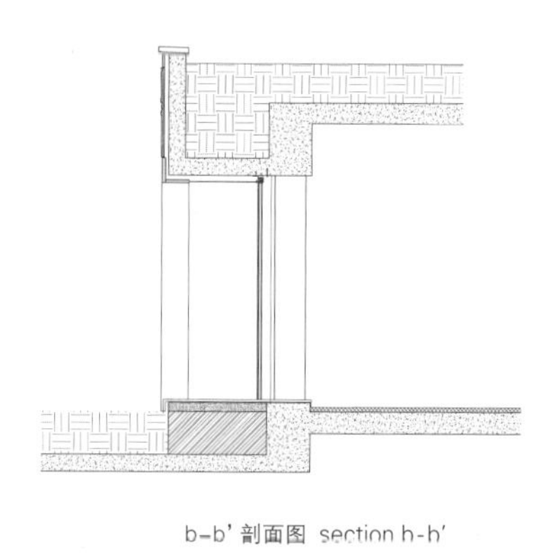
b-b' 剖面图 section b-b'

Space Planning – unifying diverse departments

Planning of the program allocates individual zones for different departments – such as watches, jewelry, eye wear and accessories – but these are connected to each other via atriums which bring in light and allow hot air to escape. These atriums house lifts and staircases. The depth of the office spaces allow them to be enveloped entirely with daylight. Where artificial lighting is required, its use is determined by light and occupancy sensors and minimized. Solar panels above the western terrace and above the service yard generate on-site energy, providing 25% of the energy requirement.

Along the slightly-raised ground floor of the building is a cascading body of water body which forms the central spine of the building.

Reflections on the water during early morning hours, creating a beautiful game of movement and rhythm, give the space an extra dimension. By lifting up the ground level of the building, by around 2m, there is a seamless connection between the internal "bio lake" and the external Veerasandra Lake, making the boundary between them disappear.

Along this "central spine" of water is a linear, double-height space, integrated with a series of wide steps, courtyards, product display walls, seating and informal meeting spaces. This culminates in an atrium which leads to the dining block, framing the greenery. Here one can dine either indoors or outdoors, listening to the sounds of birds and flowing water. The dining block is on three levels to double up as a multipurpose hall.

Atriums – connection and community

The five atriums in the campus vertically connect all the floors, from the basement upwards, bringing in natural light and creating a sense of community. The "vertical park" landscape design is integrated into the dynamic architecture, seamlessly merging the outside and inside.

Ultimately, the Titan campus reinterprets the idea of a corporate office and seats it amidst nature to inspire and elevate the experience of the end users.

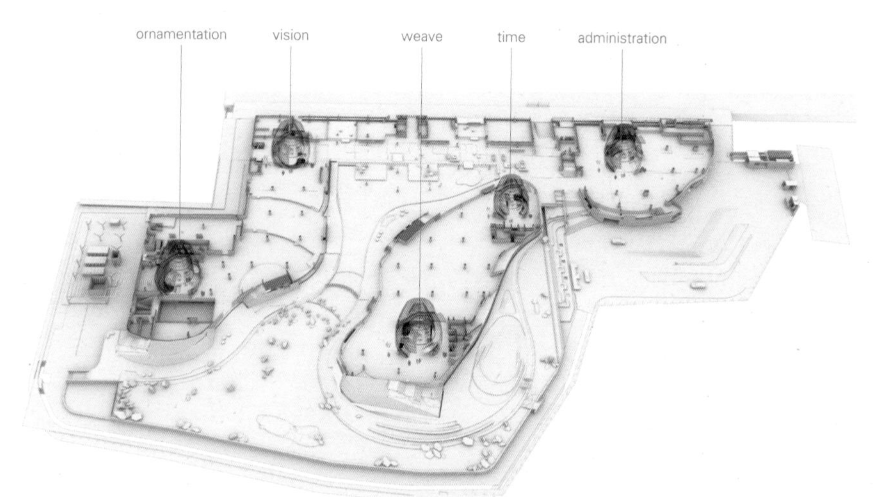

校园内的五个中庭垂直连接地下室的所有楼层，引入自然光线，同时也象征着每个部门。
The five atriums in the campus vertically connect all the floors from basement bringing in natural light and also signify each of the departments.
中庭 atriums

项目名称：Titan Integrity Campus
地点：Electronic City, Bangalore, Karnataka, India
建筑师：Mindspace
设计团队：Sanjay Mohe, Suryanarayanan.V, Swetha A, Joseph K T, Er. Mahesh.S
客户：Titan Company, Limited
结构顾问：Sterling Engineering, Bangalore
景观设计师：One Landscape, HongKong
室内设计师：MMoser, Bangalore
HVAC顾问：Airtron Consulting Engineers Pvt Ltd
PHE、消防顾问：Maple Engg-Design Services (Inda) Pvt Ltd
电气顾问：Sripeksha Engineering Consultancy Services Pvt Ltd
照明设计师：Light Vista, Bangalore
总建筑面积：36,232.19m²
竣工时间：2017
摄影师：©Purnesh Dev Nikhanj (courtesy of the architect) - p.72~73, p.74, p.76, p.77, p.79, p.82, p.84, p.85[lower]; courtesy of the architect - p.75, p.83, p.85[upper]

中庭剖面图 atrium section

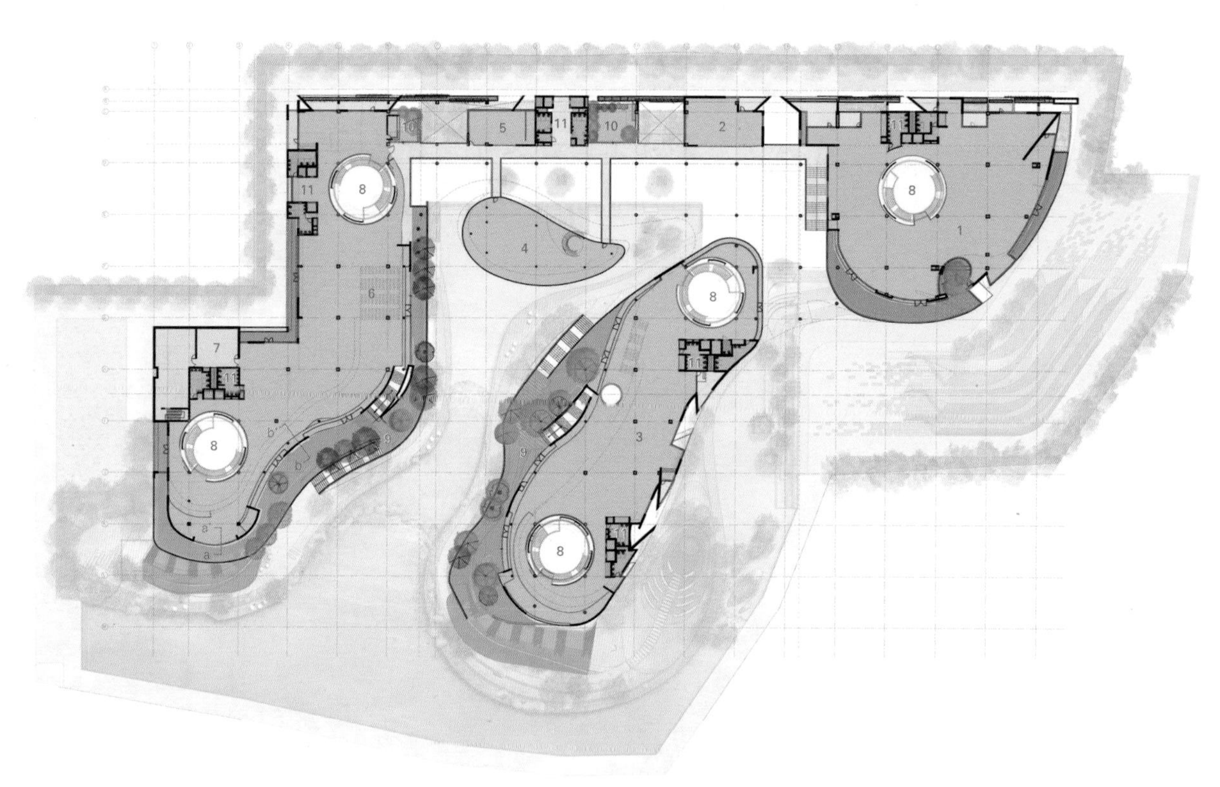

1. 财务部 2. 内部审计办公室 3. 手表部门 4. MD办公室 5. 董事会会议室 6. 眼镜部门
7. 电气室 8. 中庭/楼梯 9. 露台花园 10. 庭院 11. 电梯/卫生间
1. finance department 2. internal audit office 3. watches division 4. MD's cabin 5. board room 6. eyewear division
7. electrical room 8. atriums/staircase 9. terrace garden 10 courtyard 11. lifts/washrooms
二层 first floor

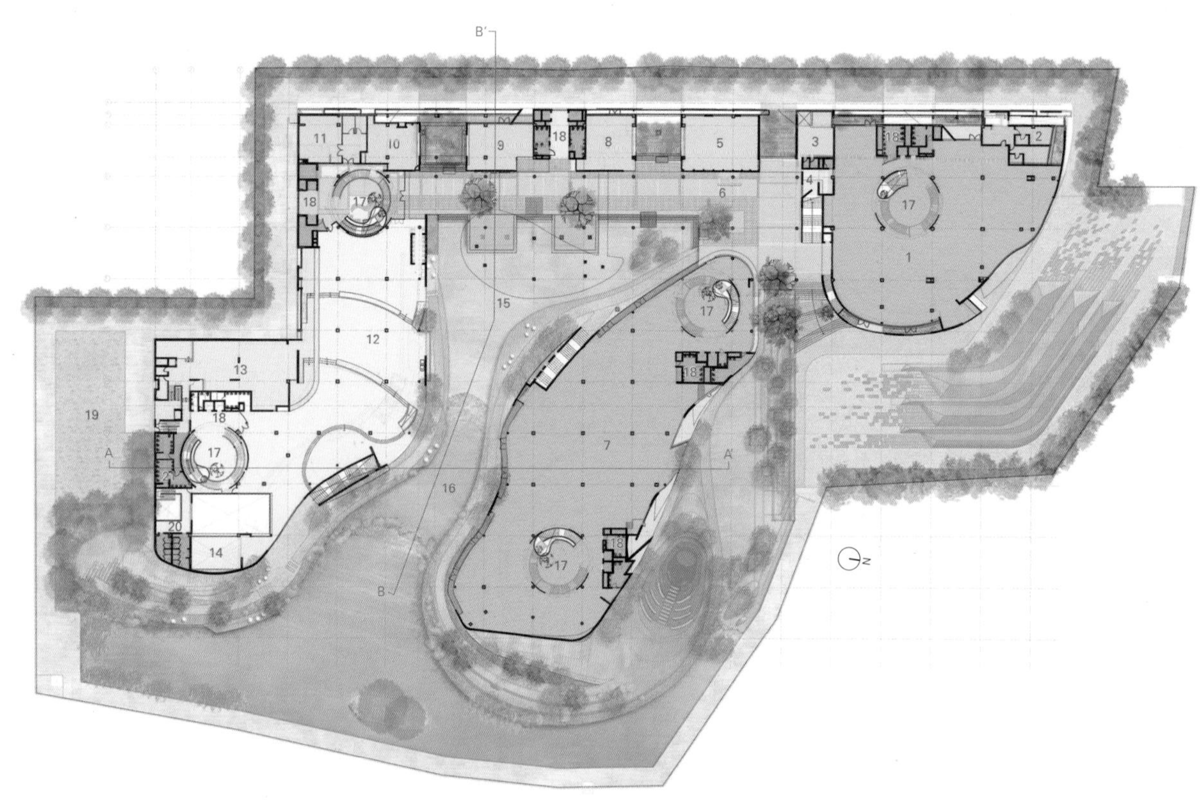

1. 管理办公室 2. 休息室 3. 托儿所 4. 医务中心 5. 培训室 6. 中央轴线、接待处、等候处 7. 手表部门 8. 讨论室
9. 产品展示间 10. 图书室 11. 服务器机房 12. 餐厅/多功能厅 13. 厨房 14. 运动场/健身室 15. 消防水箱 16. 生态湖
17. 中庭/楼梯 18. 电梯/卫生间 19. 设备间 20. 健身房更衣室/卫生间 21. 服务人员更衣室/卫生间
1. administration 2. refreshing rooms 3. crèche 4. medical center 5. training room 6. central spine, reception, waiting 7. watches division
8. discussion rooms 9. product display room 10. library 11. server room 12. dining/multipurpose hall 13. kitchen 14. sports/gym 15. fire tank
16. bio lake 17. atriums/staircase 18. lifts/washrooms 19. services 20. gym change/washrooms 21. service staff change/washrooms
一层 ground floor

1. 露台花园 2. 设备区，上方安装太阳能板 3. 采光天窗 4. ACU
1. terrace garden 2. services with solar panels above 3. skylight 4. ACU's
屋顶 roof floor

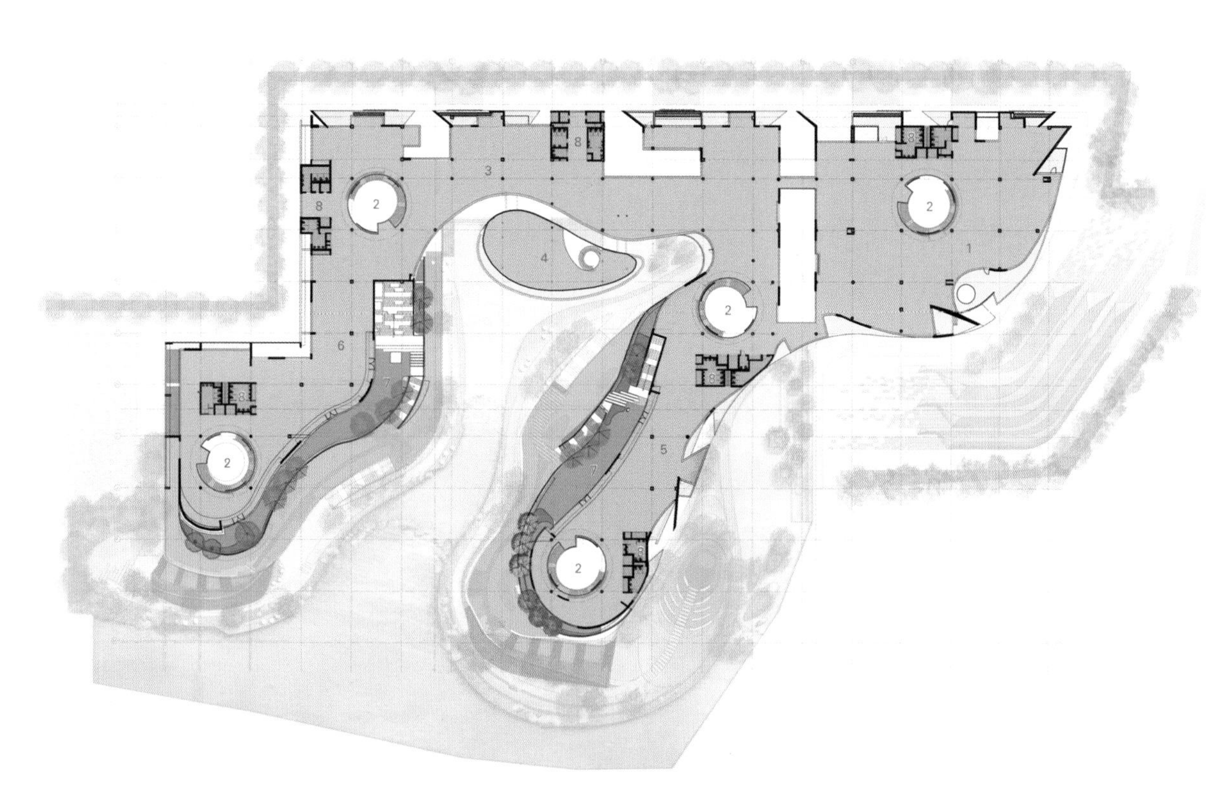

1. IT/公司可持续性研究部门 2. 中庭/楼梯 3. 珠宝部门 4. 讨论室 5. 沙丽服部门
6. 设计师/陈列师办公室 7. 露台花园 8. 电梯/卫生间
1. IT/corporate sustainability division 2. atrium/staircase 3. jewellery division 4. discussion area
5. Taneira-Saree division 6. designers/visual merchandise 7. terrace garden 8. lifts/washrooms
三层 second floor

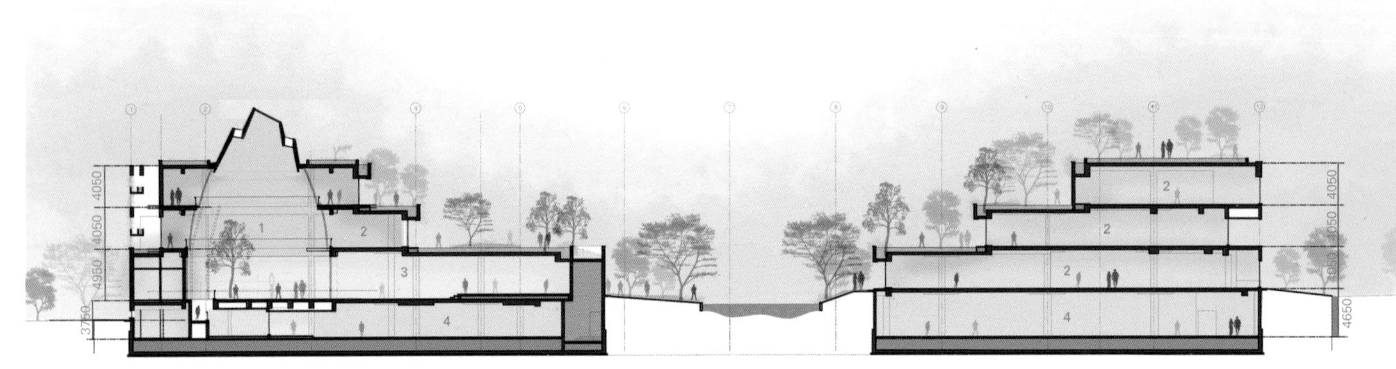

1. 中庭 2. 办公室 3. 餐厅/多功能厅 4. 停车场/设备间
1. atrium 2. office 3. dining/multipurpose hall 4. car park/services
A-A' 剖面图 section A-A'

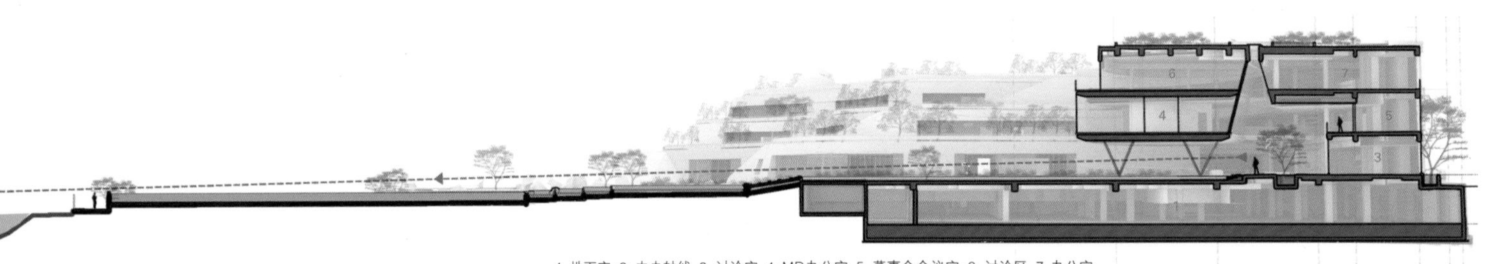

1. 地下室 2. 中央轴线 3. 讨论室 4. MD办公室 5. 董事会会议室 6. 讨论区 7. 办公室
1. basement 2. central spine 3. discussion roooms 4. MD's office 5. board room 6. discussion area 7. office
B-B' 剖面图 section B-B'

TITAN

历史价值的适应性再利用

Adaptiv with Histo

我们的城市正在以惊人的速度扩张，因此我们必须开始认真考虑现有的建筑中可能具有保存价值的那部分。如今遍地都是起重机，各种施工作业随处可见，为了适应不断增长的人口，新的大楼、新的基础设施和便利设施的建设正如同雨后春笋一般持续快速地涌现。

从本质上来讲，我们都清楚，如果不控制碳排放，像这种规模的扩建对环境的各个方面来说都是一场灾难，而且这场灾难的影响会长期地持续下去。众所周知，关于建筑和可持续性、合理材料的选择、建筑方法，特别是与新建筑、适应性再利用项目相关的建筑法规，全球都在进行激烈的讨论。

Whilst our cities grow at phenomenal speed, many more existing buildings need to be seriously considered as potential keepers. Horizon lines are already dotted with cranes, evidence of continued fast "mushrooming" construction, new building projects, new infrastructure and amenities for the ever-growing population.

Essentially, we know that the environmental factors of newbuild to this scale is disastrous, with long lasting effects, if we do not reduce the carbon impact. As we are all aware, hot discussions are taking place around the globe on architecture and sustainability, selection of justifiable materials, methods and, notably, codes and regulations related to newbuild versus adaptive reuse.

e Reuse rical Value

然而，由于种种原因，对于不再满足其初始用途的建筑来说，它们的适应性再利用已经变成了首要议题。也许因为基础设施需要更换、迁移或者淘汰？也许是公司的办公空间越来越大，而现有地段无法满足，必须搬到更宽敞、更舒适的场所？又或许是技术创新改变了机遇和商业模式？

总之，建筑趋势发生了明显的转变，那就是越来越多的建筑师正在寻找现有的一些风格有趣、历史价值突出的建筑，细心谨慎地对它们加以保存、改造和重用。通过对现有建筑的和谐再利用，我们应该将保护建筑遗产和必要的文化价值以及场所文脉放到首要议程中。

However, adaptive reuse is high on the agenda for buildings that no longer serve their original purpose, for one reason or another. Perhaps infrastructure was changed, moved or discarded, or companies and their offices simply grew too big for their location, and moved on to bigger and better places? Perhaps innovation in technology changed the opportunities and business models? There is a clear drift that shows more and more architects are seeking to work on existing, interesting, historically valuable buildings, in order to sensitively conserve, adapt and reuse. Saving architectural heritage and also imperative cultural values, and place, through sympathetic reuse should be on top of the agenda.

历史价值的适应性再利用

Adaptive Reuse with Historical Value

Heidi Saarinen

与其拆除闲置的空间和建筑物，现在的趋势是对其重新配置，提升使用效率和质量，使建筑获得新生。考虑到当前的气候，以及所谓的"用完即丢的一次性消费文化"，对现有建筑的重复利用和改造成为一种更好、更真诚的推进方式。人们对"大道至简"的理念越来越感兴趣，通过对曾经服务于其他用途的各个空间进行改造，使之满足当前的用途，为这些建筑增添了宝贵的特点和新的记忆。建筑原先古旧的特征应该突出显示出来，与新增的部分相互对照，相互协调，创造出一种非常独特的格调。[1]

在历史留下线索、痕迹和灵魂的空间里，设计理念随处可见。将这些设计理念引入到设计过程的里里外外，对用户、居民和社区以外的人来说，这是一种更为有趣的叙事方式。

通过了解建筑的历史以及它在整个使用周期内所创造的场所，无论是新建建筑，还是对老建筑的适应性再利用，都将作为该场地的历史框架发挥作用。只要现存的装饰、历史性的立面和有创意的片段可以得到保存、修复，甚至可以作为该建筑的价值宣言得到"展出"，那么它就值得人们为之奋斗。因此，这种建筑的重新使用成为一个重要的路标，提醒我们所有建筑遗产的过去。

然而，"唤醒"一栋需要重建的、闲置或被忽视的建筑会产生很多的问题。尽管拆除和重建似乎"更容易"，但通过恢复、保存和改造使它们符合当前的用途和规定，从中产生的意义将是无价的。[2]它先前独有的历史、材料和使用经历可以为项目带来的价值不可小觑。

本文探讨了五个值得关注的项目，并对项目开发中的一些相似之处和需要注意的细节进行了探讨。作者列举了它们的历史价值与当代设计的效率、技术、材料和工艺，给这些项目在旧建筑的修复、保护和使用创新上做出的努力点赞。

洛哈尔（LocHal）图书馆（108页）位于荷兰蒂尔堡市中心，紧邻一条铁路线，建筑被大型玻璃和钢结构的外壳包裹，很明显，这在主题上呼应了它过去作为铁路建筑的那段历史。该图书馆由Civic Architects建筑师事务所建造，在该地区营造了一种全新的氛围。这里曾经是交通和火车枢纽，现在是汲取知识，探索学问的地方。洛哈尔图书馆借鉴原来的结构设计，增加了新的空间、楼层、活动中心和功能区。其中一些结构为了在空间上有所区别，必要时可由大型布帘隔开，这些布帘覆盖了地板到天花板整个15m的高度，横跨整个空间，用来保护隐私、划分功能区，提供安静的空间或公共礼堂。这些由蒂尔堡纺织实验室编织的布帘同时也是巨大的声学元件。

Instead of demolishing unused space and buildings, one ongoing contemporary trend is to reconfigure buildings into a new, more efficient or better use. Considering the current climate, and what could be called "throw-away-plastic-consumer-culture", re-use and adaptiveness of existing architecture is a better and more honest way forward. There is a growing interest in "less is more" and by reinventing spaces that once had a very different use, into a contemporary use, will add invaluable character and memory to those buildings. Timeworn, original features should be highlighted against the additions, and work in harmony, creating something very special.[1]

Design ideas are plenty in spaces where history has left clues, traces and soul. To bring these into the design process inside and out will narrate more interestingly to the user, occupier and the community beyond.

By understanding the history of the building, and indeed the place that it has created over its lifespan, the new, or adapted use, will function also as a framework of the very history of that site. Wherever existing ornamentation, historic facades and snippets of originality can be saved, restored and even "exhibited" as a value statement of that building, it should be strived for. Therefore, the building's new, current use, acts as an important signpost, reminding us all of its heritage past.

However, there are many issues with "waking up" a tired, unused or neglected building. Although it may seem "easier" to demolish and rebuild, the significance added by restoring, conserving and adapting to current use and regulations will be invaluable.[2] The worth that such individuality of history, materials and previous use can add to a project cannot be underestimated.

In this text we explore 5 noteworthy projects, with some similarities between the project developments and attention to detail. Juxtaposing their historical values with contemporary design efficiency, technology, materials and processes gives these projects a thumbs up in terms of restoration, conservation and new innovative use.

Located in the Tilburg city centre, in the Netherlands, next to a railway line, within a tall industrial glass and steel shell, clearly resembling the building's railway themed past, the LocHal Library (p.108) by Civic Architects gives the area a totally new vibe. Having been the transport and train hub, this is now a space for knowledge, discovery and learning. LocHal Library has been designed around the original structure, with new additional spaces, levels, hubs and zones. Some of these are differentiated spatially and separated where necessary by large textile screens, covering the entire 15m floor to ceiling span, moving across the space for privacy, different functions or to highlight quiet or auditorium space. The textiles, woven by Tilburg Textile Lab also act as enormous acoustic elements.

The reception desk, stairs and foyer areas, nested beside the otherwise dark industrial materials and palettes, are bright,

接待处、楼梯和门厅区域明亮有趣，主题明确（接待处和一些座位是用图书馆内的书叠成的），而它们周围的设计则采用了色调完全相反的黑色工业材料。中央楼梯也可以当成座位区，用作会议区/社交区/学习区，它由混凝土、玻璃、钢材和橡木制成，让空间具有了轻快的现代氛围。沿楼梯/座位区安装了可以给笔记本电脑和其他设备充电的数据接口。该大楼还充当了一个网络中心，在这里联合办公的企业和活动组织可以在这里举办聚会。

在整个图书馆中，使用者会产生一种闯入不同空间的错觉，就像是在穿过一个城市的市集广场，人们在这里可以阅读、开会、茶歇。自然光通过令人惊叹的大型窗户进入馆内。一些绿植从下方沿着上面的楼座摆放得恰到好处，形成视觉上的冲击，创造了一个有趣的阅读环境。到了晚上，图书馆灯火通明，成为夜幕降临之后的城市指针。

另一个大型空间是联合利华新泽西州总部（154页），由Perkins and Will建筑事务所设计，在那里工作的同事也同样会有一种置身在别处的喜感；按照建筑师的说法，"就好像在曼哈顿一样"。在工作和娱乐之间的任何时间和场合，员工们都可以放松一下，沿着大厅散步，与他人交际，谈谈业务或聊聊趣事。设计师没有拆除始建于20世纪六七十年代的一些现存建筑，而是将它们相互连接起来，创造出一个大型的校园风格空间。以前的院子现在变成了室内中庭，设有休闲区和社交区。员工可以购物、买茶点、在健身房锻炼，甚至在休息时间理发。这里没有指定的办公室，而是采用一种灵活的轮用化办公制，员工可以在各种不同的地点工作。除了指定的专业空间和会议空间外，还有一些私密安静的空间和偏日常化的社交/工作空间，例如，兼作座位区的大型中央楼梯，非常适合用来应对"空降的"工作会议，或是抽出一些时间安静地跟进一下邮件通信。还有一些更为亲切的"起居室"区域，员工可以在这里工作、放松，"起居室"甚至连壁炉都有。

公司监控不同空间的使用情况，以便了解优先使用的空间和优先安排的活动，来提高空间的生产率和使用率。

此外，公司还在纽约市和新泽西州的关键地点提供班车服务。这有助于减少汽车的使用和停放问题，形成一种方便的上班方式。更重要的是，75%的材料来自垃圾填埋场。这种环保材料的使用和包容性的设计决策，为该建筑设计赢得了享有盛誉的美国绿色建筑委员会LEED白金奖。

fun and theme specific (reception and some seating were made out of stacked-up library books). The central stairs, also acting as a seating and meeting/social/study area, are made of concrete, glass, steel and oak, allowing a lighter, contemporary ambiance into the space. Electrical and data points for charging of laptops and other devices are easily accessed along the stairs/seating. The building also acts as a networking hub, where co-working businesses and events organisations can host gatherings.
There is a feel of entering different spaces within the overall library. Users are made to feel as if they are walking through a city market square with spaces for reading, meeting and a café for refreshments. Natural daylight enters the building through the tall, stunning windows. Plants and greenery are appropriately placed in the space, visually striking from below along the upper balcony levels, creating an interesting environmental reading habitat. At night the library lights up, acting as a pointer in the city after dark.
Unilever's New Jersey headquarters (p.154), designed by Perkins and Will is another vast space, where similarly its co-workers are playfully made to feel as if they are somewhere else; "[..] in Manhattan", according to the architects. Between work and play, staff can relax, walk along the concourse and mingle with others, talking business or pleasure, whatever the situation or time of day may be. Instead of demolishing several existing buildings from the 1960's and 1970's, the buildings were linked to create one large campus style space. A former courtyard now acts as a central indoor atrium; with relaxing- and social areas. Employees can shop, get refreshments, workout in the gym or even have a haircut in their breaktime. Without assigned desk space, this is a flexible, hot desking office, allowing working from a range of different settings. Alongside designated professional and meeting spaces, there are some intimate, quiet spaces and more informal social/workspaces, like the large central staircase that doubles as a seating area, perfect for "accidental" workspace meetings and maybe some quiet time catching up on emails. More familiar environments can be found in the "living room" areas, where staff can work or relax, there is even a space with a fireplace.
The company monitors the use of the different spaces, so there is an understanding of preferred spaces and activities, to improve the productivity and use of space.
Furthermore, a company shuttle service is offered from New York City and key locations in New Jersey. This helps reduce car use, parking issues and is a convenient way to get to work. More importantly, 75% of the materials used were derived from landfill. This environmental use of materials and inclusive design decision making, earned the company the prestigious US Green Building Council award LEED Platinum.
After reorganisation at the advertising agency's BBDO's Brussels office (p.142), in Sint-Jans-Molenbeek, Belgium, ZAmpone

在比利时布鲁塞尔的莫伦贝克-圣让，BBDO公司（142页）进行重组后，委托赞普建筑事务所做出概念上有所创新的设计，要求既能保留1908年公司最初上市时的建筑元素，同时又具有现代的建筑风格。这座建筑曾经是葡萄酒的仓库和贸易地点，后来经历了一系列的改造；从1999年开始，最初的仓库被改造成了BBDO办公室。赞普建筑事务所在2017年完成了对现有建筑的最新翻新和扩建，包括开放式的空间布局、为员工设计的中庭、休闲区、采光良好的轮用制办公区；可眺望远处风景、帮助激活创造力的视觉设计；可以将社交和商业活动紧密结合的空间安排，如餐厅和流通区。当会议在特定区域举行时，噪声的音量会被控制到最低。会议空间没有窗口，这就使得这些空间可以得到有效的使用。建筑师时刻注意要在可靠的预算范围内充分利用空间。空间活力、声学设计和自然光的利用是该设计的核心概念。

为了适应BBDO的文化，空间保持了流动性，传统的固定办公室和工作站被换成了灵活的空间。因此，空间是共享的，在不同的时间以不同的方式被使用，从而有助于解决每天办公室里的各种实际问题和财务问题。建筑一层是与其他公司共享的，为专业和社交的合作制造机会，费用共同分担。

在空间上，实行巧妙的分区和交通流线设计，如迁移主入口，使它与停车场连接，这一决定从根本上改变了空间的活力，和以前的空间安排有所区别。此外，该设计使得大厅这个巨大的空间与各处交汇，明确定义它的核心地位。从外部可以观察到内部的活动和运转情况，让我们可以看到这个创意"引擎"的日常生活。并置的形式介入并分隔了主要空间，如藤蔓一样的彩色移动式钢管里面包含电线和数据线，可以巧妙地根据用途改变形状，比如，可以用作这个灵活工作区域的衣帽架和隔声屏，这里的一切都富有能动性、交互性和进步性。

对传统细节的另一个关注点是对前里昂工厂吉拉德厅，又称H7（92页）的翻修，Vurpas Architectes建筑师事务所对它进行了精心的改造。原来的一座建于19世纪的锅炉厂如今强烈地反映了它的传统，并且展示了其美丽的翻新后的工业细节，还有现代元素、装置、配件以及建筑材料。由于该建筑是对其过去的重要叙述，建筑师敏锐地恢复并调整了这些空间，把对建筑原始结构的干扰降到最小，材料也选用得十分恰当。吉拉德厅位于康弗伦斯区，那里是一个重要的商业区，也是欧洲最大的城市中心开发区之一。康弗伦

Architectuur were instructed to work on innovative concepts that would appreciate the original, listed building, built in 1908, yet in a new contemporary way. Once a wine warehouse and trading point the building underwent a series of consequent remodelling phases; starting in 1999 when the original warehouse was converted into the BBDO office. ZAmpone Architectuur completed the latest refurbishment and additions to the existing building in 2017, including open plan spaces; atriums, relaxation and rest spaces for employees, hot desking zones with very good daylighting; views out to the landscape beyond, helping creativity, and more intimate spaces where social and business interaction coexists, such as in the restaurant and around the circulation areas. Noise is kept to a minimum as meetings are held in specific zones. The meeting spaces do not have windows, allowing the spaces to be used efficiently. The architects were mindful of using space resourcefully and within a credible budget. The key conceptual components were dynamics, acoustics and daylight.

Adapting to the culture at BBDO, the space remains fluid, as traditional fixed offices and workstations have been exchanged for flexible spaces. Therefore, space is shared, utilised in many different ways, at different times; helping the practical and financial side of the office day-to-day. Ground floor areas are shared with other companies, creating professional and social collaboration opportunities and sharing of expenses.

Spatially, there are clever zoning and circulation decisions, such as moving the main entrance so it could link with the car park, changing the dynamics of the space radically from the previous arrangement. Furthermore, this allows the main hall, a massive space, to become a mingling and through fare, clearly defining this as the heart of the building. Interior activities and movement can be observed from the outside, allowing a glimpse into the everyday of this creative "engine". Juxtaposed forms intervene and separate the main spaces; colourful mobile trailing steel tubes containing electrical and data cables, cleverly transforming into other uses such as coat stands and acoustic screens in the flexible workstation areas. Everything here is about dynamics; interaction and moving forward.

Another attention to heritage detail was followed in the renovation of former Lyon factory, Halle Girard or H7 (p.92). It has been mindfully redesigned by Vurpas Architectes. A 19th-century former boiler shop today strongly echoes its heritage; showing off its beautifully restored industrial details alongside contemporary elements, fixtures, fittings and materials. Because of the building's important narrative of its past, the architects sensitively restored and adapted the spaces with minimal interference to the building's original fabric. Appropriate materials have been used throughout the project. Located in La Confluence, a key commerce area, and one of the largest city centre developments in Europe, where environmental, technical and architectural innovation is strongly encouraged.[3]

Striking external features and interiors fit into Halle Girard's location between the city buzz and the landscape beyond. The atmosphere in the main industrial hall space is "cathedral like", according to the architects, with its vast hall opening

斯区的大力鼓励环境、技术和建筑方面的创新。[3]

吉拉德厅的外观和内景都十分引人注目，与其介于闹市区和外部景观区之间的位置特点十分契合。根据建筑师的说法，主要工业大厅的空间氛围是“大教堂式的”，其巨大的空间向远处的景观开放。该建筑已改造成为数字创新创业企业的中心，吸引居民来此寻找新的工作模式。为了配合周边地区的发展，这一改造经过了精心规划，把智能技术和未来远景均考虑在内。

室外光线一方面通过一层的大型落地窗进入室内，洒在用于放松和举行临时会议的休闲座位区，另一方面又从上方的屋顶射入较小的棚屋空间。室外只有雅致、简单的坐席区和绿化区，没有附设多余的装饰物，以此来突出建筑本体。

由Valentiny hvp Architects建筑师事务所设计的卢森堡大学学习中心（126页），也许是本章最宏大的项目，不仅因为其使用目的发生了极大的变化，而且建筑造型也完全改变。最初的框架以许多复杂的方式被保存、加固、清洁和增扩，用于纪念曾作为工作筒仓的工业历史，与建筑的新用途和许多新的扩建结构形成了对比。这些大型的创新性开放空间可以用于学习、阅读和举办活动，其中一部分区域也向校外公众开放。这些空间分布在五个楼层，复杂的六边形结构和整个建筑的外框给人以运动和交流的印象。室内布局和照明都经过了精心的设计，可以满足社交活动和人流需求。此外，窗户玻璃有的透明，有的不透明，使工作和阅读区域获得适当的照明。有趣的是，有些玻璃专门进行了丝网印刷制作，让立面从外侧呈现出一种“透气”效果。为了实现这项工程，建筑师从造船业借鉴了整体施工技术。

在研究了这些项目之后，很明显可以看出，对建筑进行适应性再利用的确具有真正的价值。作为建筑师或设计师，他们要搜集如此多的、真实而又重要的历史资料用于研究，并将之转化成概念构型，以这种方式做出来的工程无疑要比快速修复高层建筑更有效，也更有趣。当然，在这样的时代，建筑设计的方方面面都存在许多复杂的问题。因此更重要的是，尽可能以一种真正深思熟虑的方式，在国家和全球范围内，继续珍视、修复和保存我们过去的建筑和环境。现在已经太晚了，作为一个集体，我们必须更加努力地工作，尽可能为下一代传承一个体面的未来。建筑师和设计师们，你们是决策者，环境的警钟已经大声而清楚地敲响了。

up to the landscape beyond. The building has transformed into a hub of innovative digital start-up businesses, attracting inhabitants for new working models within this location specific space. The renovation, as the city developments around it, has been carefully planned with smart technology and the future in mind.

Light floods in through large floor to ceiling windows on the ground floor, where casual seating areas are scattered for relaxation, and ad hoc meetings. Light also comes into the smaller shed spaces from the roofs above. Tasteful, simple outdoor seating, topography and vegetation add character to the no-frills design, allowing the building stand out in its own right.

Perhaps the most ambitious project here, Luxembourg University Learning Center (p.126) by Valentiny hvp Architects, not only because of its drastic change of use, but also the total change of form. The original framework has been preserved, strengthened, cleaned and added to, in many complex ways, celebrating the building's industrial past as working silos against the new use and many insertions. Large, innovative open spaces, some also accessible to the public, were created for study, reading and activity areas. All this is constructed over five levels, with complex hexagonal structures and overall building envelope giving impressions of movement and exchange. Interiors and lighting have been carefully planned to coincide with the circulation and footfall. Furthermore, window glass is part transparent and part opaque, allowing suitable levels of lighting into the working and reading areas. Interestingly, some of the glass was specially screen printed, giving the facade a "breathable" effect when seen from the outside. To make this project possible, overall construction techniques were borrowed from the ship building industry.

Having studied these projects, it is clear that there is real value to adaptive reuse. As an architect or designer, being able to have so much real and important history and perhaps even personal stories for research, to translate this into the forming of concepts, surely makes a stronger and more interesting project than a quick fix high rise. Naturally, there are many complicated issues in every aspect of architecture in times like these. It is even more important, therefore, to continue to value, restore and preserve, where possible, in a truly considered way – and celebrate the past in our architecture and built environment; nationally, and globally. As it is almost too late, as a collective, we must work even harder, to pass on as decent a future as we can for the next generations. Architects and designers – policymakers; environmental alarm bells are ringing loud and clear.

1. L. Wong, *Adaptive Reuse: Extending the Lives of Buildings.* (Basel: Birkhäuser, 2017)
2. W. Hurst, *More top architects throw their weight behind AJ RetroFirst campaign* [online] available from https://www.architectsjournal.co.uk/news/more-top-architects-throw-their-weight-behind-aj-retrofirst-campaign/10045132.article> [25 November 2019]
3. Urban Project (n.d.) *A Step-by-Step Approach* [online] available from <http://www.lyon-confluence.fr/en/urban-project/urban-redevelopment/> [25 November 2019]

吉拉德厅的改造
Halle Girard Building Renovation

Vurpas Architectes

H7——法国里昂数字媒体公司新的温床——在经过改造的工业环境中为初创的数字媒体公司创造了活动空间和基地

H7 – Lyon's new incubator for digital businesses – creates a space for events and a base for digital start-ups in a renovated industrial setting

康弗伦斯是欧洲最大的城镇中心扩建项目之一，占地150万平方米。20世纪90年代，它还是罗纳河和索恩河的交汇处一个被遗弃的工业荒地，而现在，重建区从市中心向外延伸，并为多达30000位新居民提供了居住场所。

吉拉德厅（H7）位于康弗伦斯区的尽头，将工业遗产与创新建筑相结合，为城市复兴模式做出了贡献。城市规划、建筑以及与自然和创造力的结合，都成了该地区的特色。

历史遗留建筑

建于1857年的吉拉德厅工厂和哈雷橡胶工厂都是该地区工业历史最后的遗迹。令人印象深刻的是其独特的19世纪工业建筑特点：装饰艺术风格的立面俯瞰着罗纳河，主楼规模壮观。该建筑位于赫尔佐格和德·梅隆设计的康弗伦斯"商业2区"布局中的一个战略位置上，位于密集城市发展和场地南端被称为"领域"的大型自然区域之间的交界处。

新的居所

H7项目对工业遗迹进行改造，以服务于新的工作方式，鼓励在该地区开展创意产业。将原来的工厂面向自然环境开放，是一种重新创造生活空间的方式，也是一种以更自由、更舒适、更透气的方式利用大型工业空间的方式。

低干涉改造模式

在不影响大型工业大厅整体性、大教堂般的体量、细长的金属框架和柔和的灯光这些内在特质的情况下，如何将新的创新工作融入到前锅炉房的历史墙体中，同时又不冲淡或扭曲历史事实？答案是严格

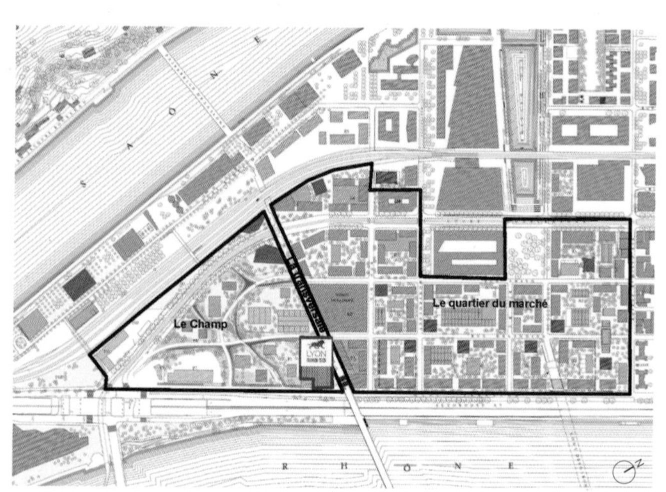

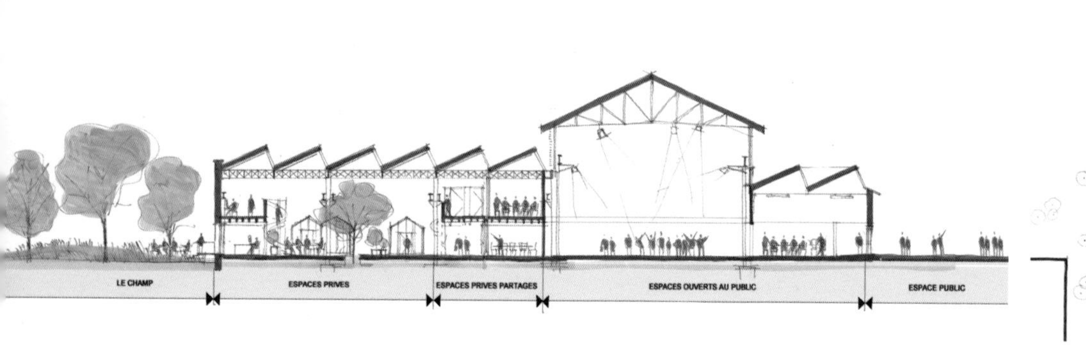

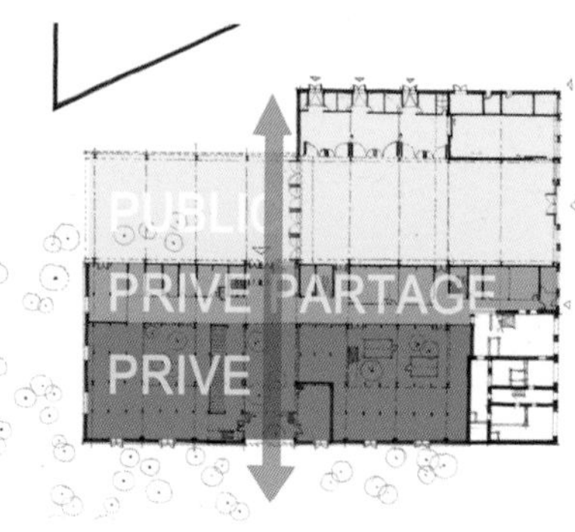

地将干预程度降到最低。作为空间的主要组成部分，设计保留了结构和外围护结构，提供了保护和与外部接触的手段。这种介入也表达了当地的建筑语言，与它的时间性产生共鸣。新的设计提出了一个简单有效的建筑应对方案，以符合工业大厅的特点和潜力，以及未来可能出现的创新性或者出现新的用途。

适应翻新的工业建筑

本项目的功能定位是明确区分两个主要空间：工作空间和活动空间。这种区分是通过优化利用现有的大厅及其空间的特殊性而得出的：空间很宽敞，活动区的主中堂没有承重的特征，而南面的大面积棚顶和高质量的顶棚采光则构成了第三空间。这种空间组织还根据所需的保密程度和居住空间的所有权，定义了一些功能分区——城市一侧的公共区域越多，面向"领域"的一侧的私人区域就越多。

把人们聚集在一起的空地

大厅的主中庭和屋顶的结构被保留了下来，形成一个巨大的空旷空间，用于抵御恶劣的天气。由城市规划师赫尔佐格和德·梅隆设计的横断面，最终将罗纳河和索恩河连接起来。横断面可以扩张，并入大厅腾出的空间，形成一个被遮盖的广场。随着能见度的提高，H7成了一个象征性的参照点，整个康弗伦斯区可以在这里汇聚成里昂数字生态系统的城市展示中心。

Confluence is one of Europe's largest town center expansion projects. Covering 1,500,000m^2, in the 1990s it was a derelict industrial wasteland at the confluence of the Rhône and Saône rivers. Now, the redeveloped area extends from the core of the city center, offering habitable space to 30,000 new inhabitants. Situated at the extremity of the Confluence district, the Halle Girard (H7) contributes to the urban regeneration model, combining industrial heritage with innovative construction. Urban planning, architecture, integration with nature and creativity define the character of this district.

A Historic Building

A former factory dating from 1857, the Halle Girard is the last vestige, along with the Halle Caoutchouc, of the area's industrial past. Its characteristic 19th-century industrial architecture makes an impressive statement: the Art Deco facade overlooks

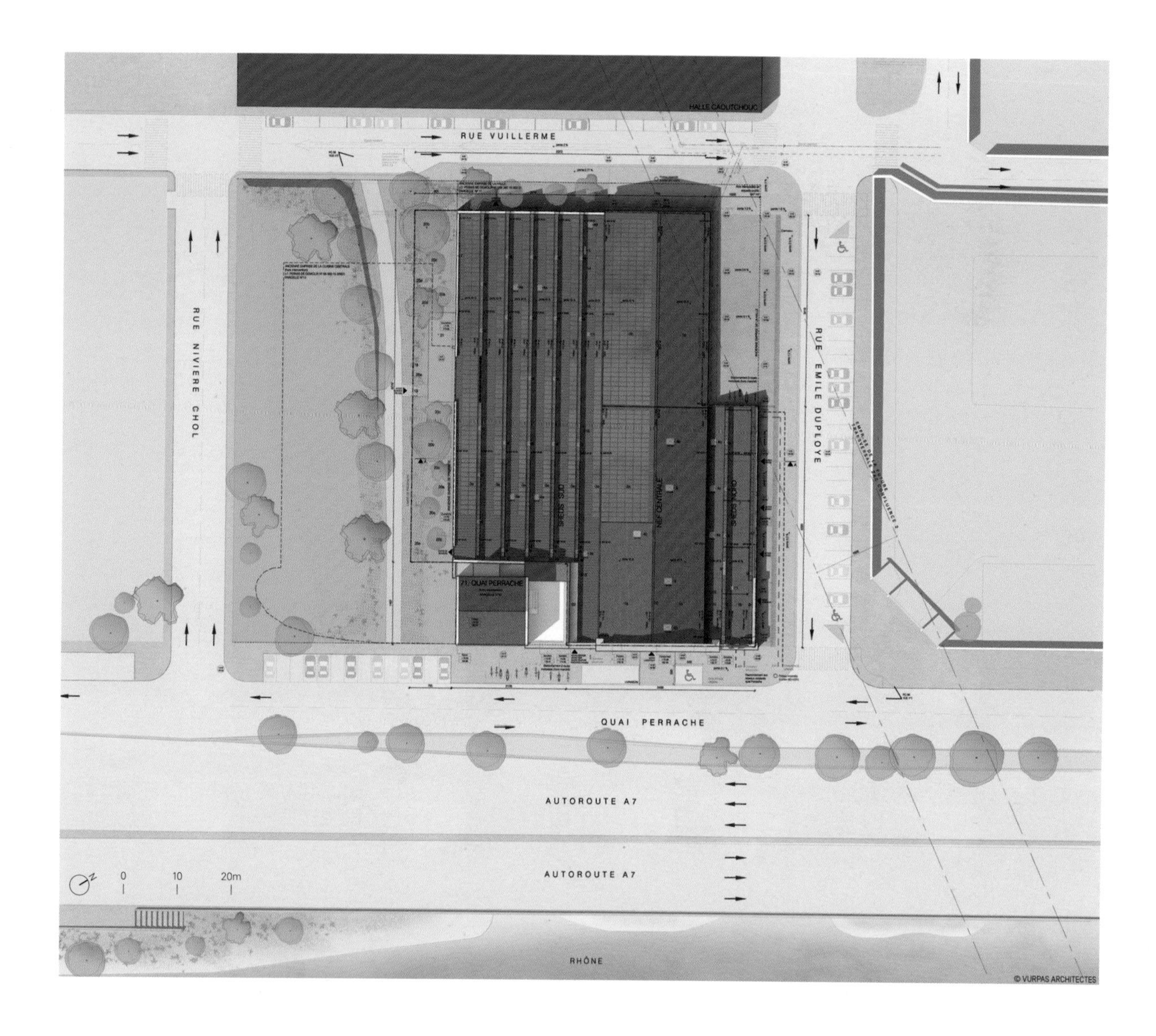

东立面 east elevation

西立面 west elevation

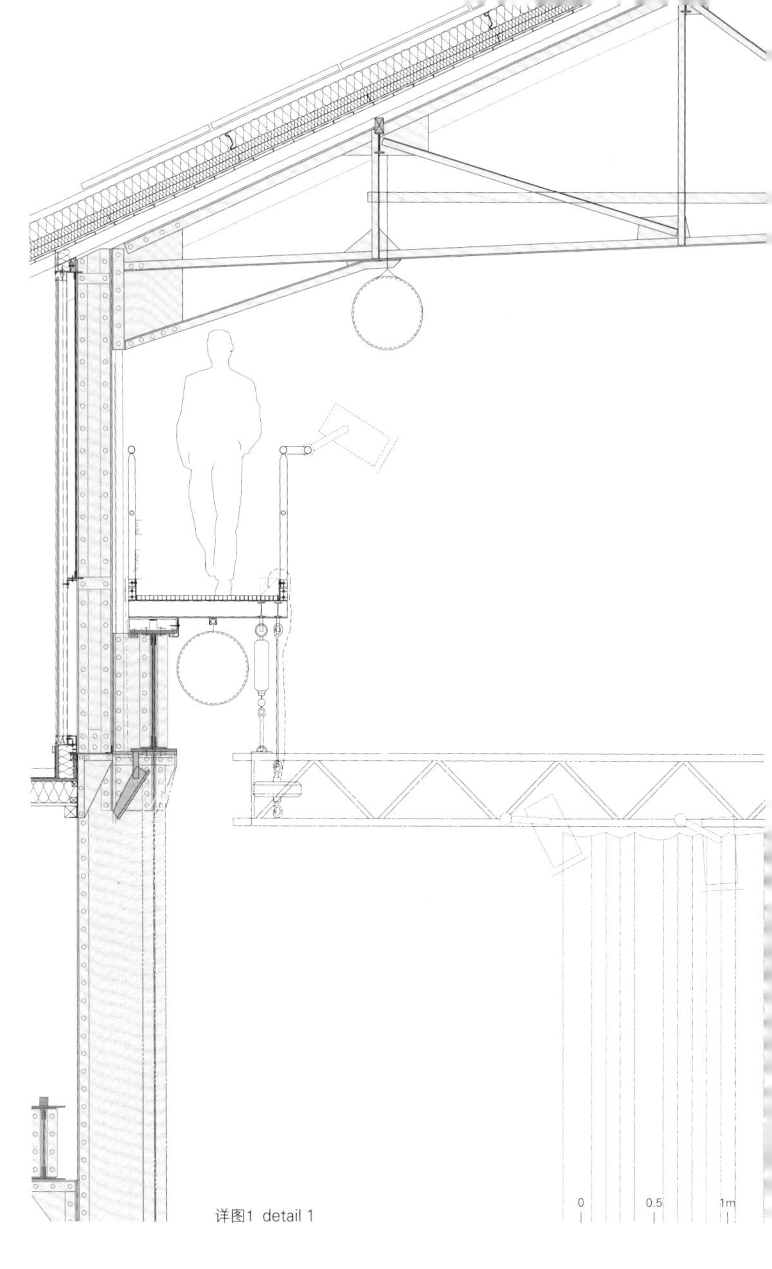

详图1 detail 1

北立面 north elevation

南立面 south elevation

COME
AND
MEET
US
Hej
Hello
Hola

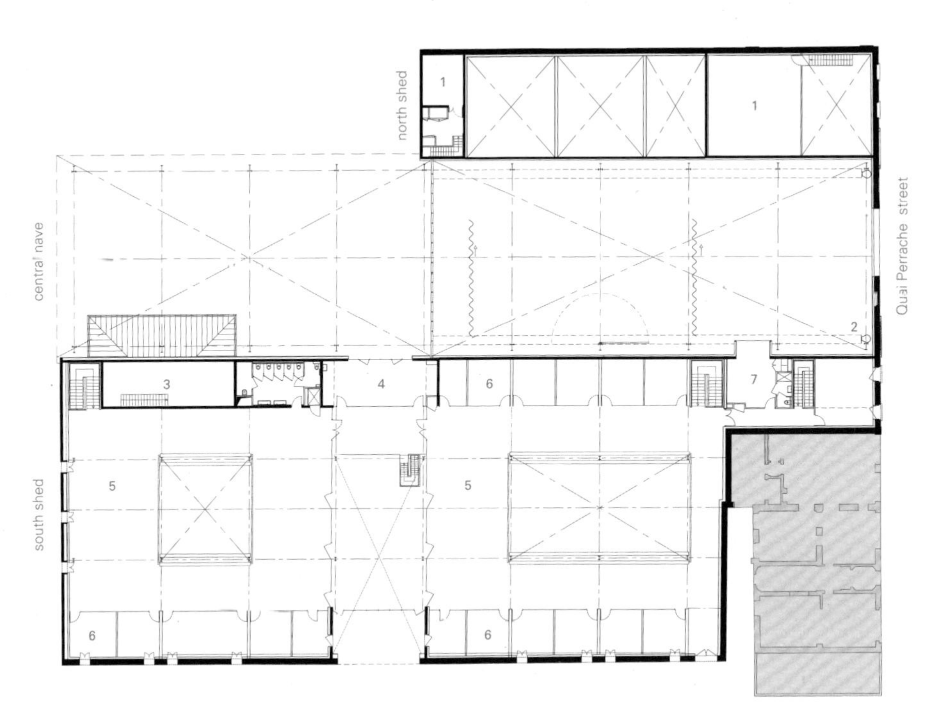

1. 技术设施 2. 技术人员通道 3. 孵化器厨房的技术设施 4. 创意房间
5. 开放空间 6. 办公室/车间办公室/会议室 7. 可以俯瞰中殿的午睡室
1. technical facilities 2. walkway for technicians 3. technical facilities for incubator kitchen 4. room for creativity
5. open space 6. offices/workshop office/meeting rooms 7. nap room with view over the central nave
二层 second floor

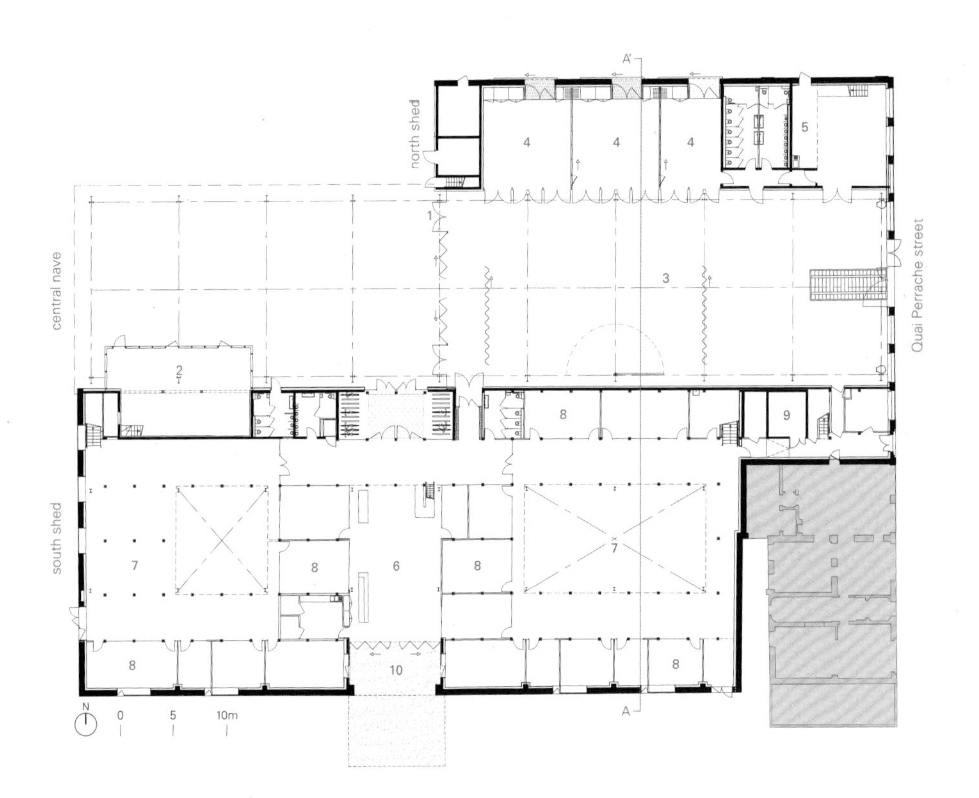

1. 食品大厅 2. 孵化器厨房 3. 活动空间 4. 小活动厅 5. 本地物流 6. 社区大厅 7. 开放空间
8. 办公室/车间办公室/会议室 9. 技术设施 10. 可以俯瞰树林的露天咖啡座
1. food hall 2. incubator kitchen 3. event space 4. smaller hall for event 5. local logistics 6. community hall 7. open space
8. offices/workshop office/meeting rooms 9. technical facilities 10. a large terrasse with view over the trees area
一层 first floor

THE START OF SOME THING
COME AND MEET US
THE START OF SOME THING
Jury n°1

the Rhône and the main building is of spectacular dimensions. The building is located at a strategic point in the layout of the Confluence "Commercial Zone 2", as designed by Herzog and De Meuron, at the interface between the development of the dense city and the large natural expanse known as "The Field" at the southern tip of the site.

A New Place to Live

The H7 project constitutes the transformation of this industrial heritage to service new ways of working, encouraging creative industries in the area. Opening the former factory onto the environment is a way of reinventing living space, and of occupying the large industrial space with more freedom, more comfort, and more room to breathe.

A Light-touch Intervention

How can new innovative workspaces be integrated into the historic walls of a former boiler shop, without affecting the intrinsic qualities of the large industrial hall – its unity, its cathedral-like volumes, its slender metal framework and its soft lighting – without diluting or distorting its history? By keeping intervention to a strict minimum. The design preserves the structure and envelope as the primary components of the space, providing protection and a means of exchange with the exterior. The intervention also speaks the architectural language of the place, resonating with its temporality. The new design proposes a simple and effective architectural response, adapted to the specificities and potential of the industrial hall, as well as the innovative, unpredictable future uses which may emerge.

Adapting to a Renovated Industrial Building

The project's functional standpoint is to make a clear distinction between the two main spaces: workspaces and events spaces. This distinction is drawn by making optimal use of the existing hall and its spatial specificities: generous volumes, with no load-bearing features in the main central nave for the events area, and large surfaces and good quality overhead light from the southern shed roofs for the tertiary spaces. The spatial organization also defines a number of functional subspaces depending on the level of confidentiality required and the ownership of the living spaces – the more public areas on the city side, the more private areas on the side facing "The Field".

An Empty Space which Brings People Together

The structure of the main nave of the hall and its roof are preserved to make a large empty space, protected from bad weather. Here, the transverse designed by urban planners Herzog and De Meuron, which will ultimately link the Saône to the Rhône, can dilate, absorbing the space cleared by the hall to create a covered square. With increased visibility, H7 constitutes an emblematic point of reference, a place where the whole Confluence district can converge in an urban showcase for Lyon's digital ecosystem.

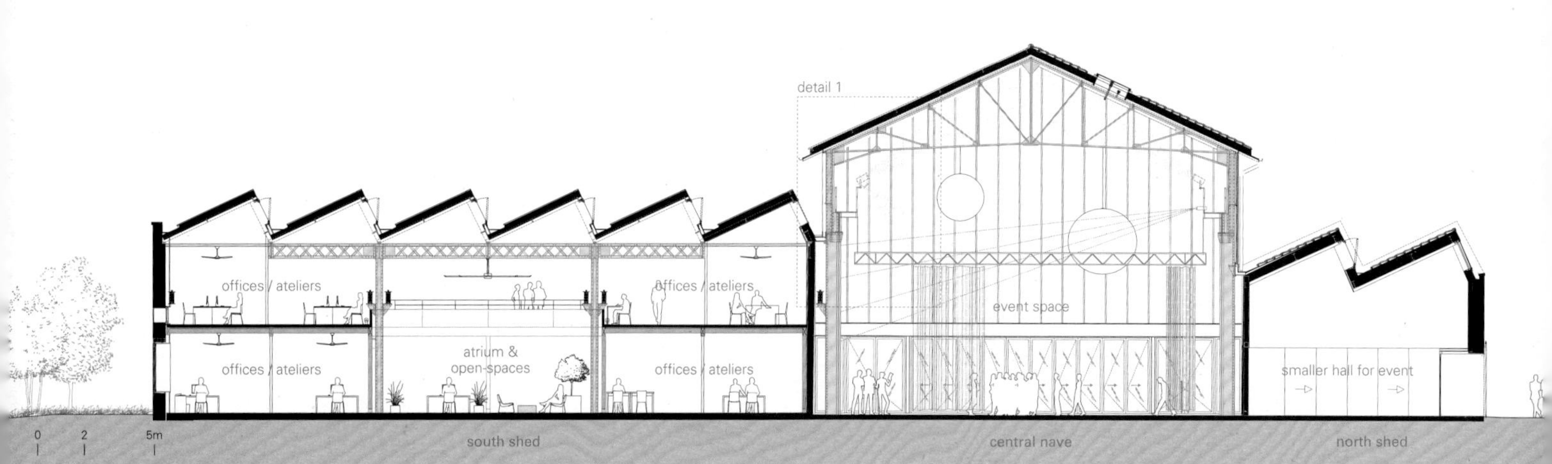

A-A' 剖面图 section A-A'

项目名称：Halle Girard / 地点：70 Quai Perrache – 69002 Lyon, France / 建筑师：Vurpas Architectes / HVAC、电力、SSI结构协调员（消防系统）：AIA Ingénierie / 声学：Génie Acoustique / 环境质量认证：AIA Studio Environnement / 客户：SPL Lyon Confluence / 总建筑面积：4,083m² / 造价：€ 7.06 million excluding tax / 设计竞赛时间：2015.9 施工时间：2017.7—2018.12 / 交付时间：2019.1 / 摄影师：©Kevin Dolmaire (courtesy of the architect) - p.97, p.98, p.99, p.101, p.103, p.104~105, p.106, p.107upper; ©Brice Robert (courtesy of the architect) - p.93, p.94~95, p.107lower

荷兰洛哈尔图书馆

LocHal Library

Civic Architects

洛哈尔，从工业遗产的巨人到公共图书馆

LocHal, a colossus of industrial heritage becomes a public library

荷兰蒂尔堡新建成的公共图书馆——洛哈尔图书馆（the LocHal）——于2019年1月在现代化的新站区正式开放。原来这里是机车车棚，进行了密集的重新设计后，具有了鲜明的铁路主题。这座优雅的工业建筑大部分被保留了下来，加上坚固的新建筑和巨大的挡帘，它已转变成一个不仅"消耗"知识而且产生知识的地方。

一个公众集会场所

新洛哈尔图书馆占地面积5400m²（90m×60m），高15m，气势恢宏，引人入胜。入口大厅是一个带顶的广场，内部有大型公众阅览桌、展览区和咖啡亭；宽阔的台阶为1000多名读者提供座位。在二层，游客在浏览书架或使用安静的阅读区时，能看到画廊和楼梯上历史悠久的玻璃墙，再往上走一层楼是一个大阳台，可欣赏到城市全景。

面向21世纪的图书馆

洛哈尔重新定义了图书馆在当今数字时代的功能。"策展"已经变得和阅读书籍一样重要；与人类专家的互动为获取知识提供了更丰富的途径。这个新的功能是由建筑学实现的；除了用于讲座和公共活动的区域外，整座建筑中还有许多技能"实验室"，如食品实验室、文字实验室、数字实验室和传统实验室。

耐用的装饰

主体结构和各种原始特征都得以精心的保护，新增加的部分使用了像黑钢、混凝土、玻璃和木材等"坚固可靠"的材料。白天，楼层、柱子和楼梯都显示出其特有的质感和缺陷。天黑后，大楼内外发生了逆转；内部成为主要的光线来源，就像城市中心的灯塔。这时，室内就有了剧院的特色，沐浴在温暖的光辉中。

挡帘

对于规模较小或供私人使用的空间，Inside Outside公司设计了六张巨大的、跟天花板等高的挡帘，这突出了建筑的规模并增强了隔声效果。这些挡帘可以通过计算机系统重新定位。当挡帘位于朝南的窗户前面时，可以减弱光线。当阳光照射到挡帘透明表面时，挡帘会变成高高的瀑布，成为宽敞的室内景观空间的组成部分。

交织的建筑设计

新的设计是对原先建于1932年的建筑的现代重新诠释。倾斜的阶

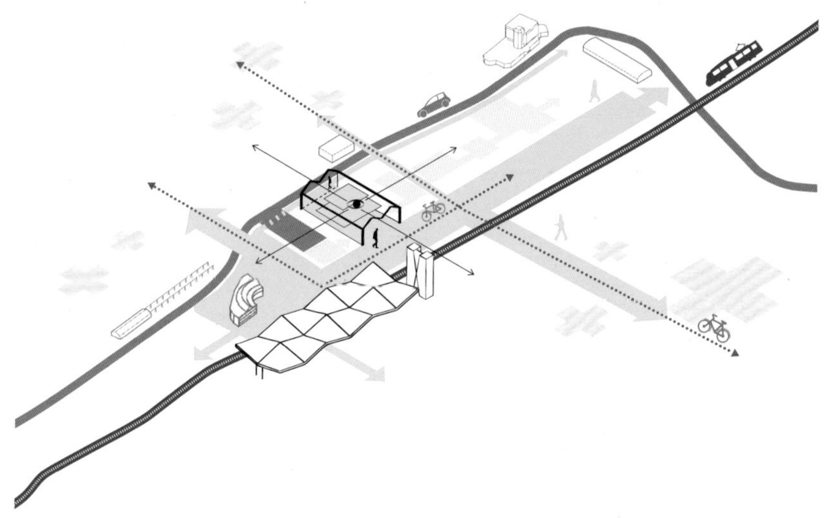

梯式室内设计在视觉上增强了空间感，给人感觉更宽敞。柱子、地板、栏杆和挡帘构成了不同轴线的"大制作"。保留下来的原始铆钉柱与大型支撑结构以更抽象的形式连接在一起，勾勒出了过道。

室内设计风格的变化

室内设计采用了很多独特的概念来定义不同的功能。橡木、钢材以及红色和橙色调与原始特征相互呼应。三张巨大的桌子仿佛火车的底盘立在原始的轨道上，形成了酒吧的延伸部分，这里也可以成为一个舞台或走秀T台。巨大的工业柱子经过改造变成了阅读和研究区域，增加了木桌和照明。图书馆的藏书放在类似书店的可移动书架上，在儿童图书馆，小游客们可以一边在浩如烟海的故事书中漫步，一边观看铅笔和尺子形状的书架。

自适应技术

洛哈尔图书馆采用分期开发方法，仅在必要时才对现有的建筑物进行加建。最终的结构采用了最初用来支撑重型机车的原有楼板和横梁来提供全部的承载力。

在一个空间里，人工气候控制的程度取决于其功能，采用了"温暖人而不是温暖空间"的原则。由五个独立的气候区组成的系统可确保最大程度地减少物理适应性，从而使洛哈尔图书馆变成一个巨大的实用体量。

Tilburg's new public library, the LocHal, officially opened in January 2019 in the newly modernized station district. The former locomotive shed underwent an intensive redesign, with a distinctive railway theme. Much of the elegant industrial building has been conserved, and with the addition of robust new architecture and huge textile screens, it has been transformed into a place in which knowledge is not only "consumed" but produced.

A Public Meeting Place

With a footprint of 90 x 60 m and a height of 15 m, the new LocHal is imposing and inviting, yet also accessible. The entrance hall is a covered square with large public reading tables, an exhibition area and coffee kiosk; broad steps become event seating for over a thousand spectators. On the second floor, the gallery and stairways reveal the historic glass walls, as visitors browse the bookcases or use the quiet reading areas. One floor higher is a large balcony offering panoramic views of the city.

A Library for the 21st Century

The LocHal has redefined the function of a library in today's

digital era. "Curatorship" has become just as important as accessing books: interaction with human experts offers a richer way to acquire knowledge. This new role is facilitated by the architecture; in addition to areas for lectures and public events, there are a number of skills "labs" throughout the building, such as the Food Lab, the Word Lab, the DigiLab and the Heritage Lab.

A Robust Décor

The main structure and various original features have been carefully preserved, and new additions make use of "honest" materials such as black steel, concrete, glass and wood. Floors, columns and stairways reveal their characteristic textures and imperfections, especially in daylight. After dark, the building turns "inside out"; the interior becomes the main source of light, like a beacon in the city center. The interior then takes on the character of a theatre, with surfaces bathed in a warm glow.

Textile Screens

For smaller-scale or private functions, Inside Outside has designed six huge, ceiling-height textile screens, which accentuate the scale of the building and improve acoustics. These can be repositioned using a computerized system. When positioned in front of the south-facing windows, the screens soften the light. As the sun hits their transparent surfaces, the curtains turn into tall cascades, becoming integral parts of the spacious interior landscape.

Interwoven Architecture

The new design is a contemporary reinterpretation of the original 1932 building. The impression of spaciousness is strength-

ened by new sightlines across the diagonally stepped interior. Columns, floors, balustrades and the textile screens form the "grand gestures" of the various axes. The original, preserved riveted columns are joined by large supporting structures in a more abstract form which delineate the aisles.

Variation in Interior Style

The interior design employs a number of distinct concepts to define different functions. Characteristic original features have been combined with oak, steel and a palette of red and orange hues. Three huge tables, like a train's undercarriage, stand on the original tracks, forming an extension to the bar that can also become a stage or catwalk. The immense industrial columns have been repurposed as reading and study areas by the addition of wooden tables and lighting. The library's collection is arranged on movable bookcases, akin to a bookstore, whilst a children's library allows young visitors to wander through giant storybooks and browse bookcases shaped as pencils and rulers.

Adaptive Technology

The technical aspects of LocHal are based on a phased approach, with additions to the existing building only introduced where necessary. The resulting structure uses the full load-bearing strength of the original floors and beams, originally designed to support heavy locomotives.

The degree of artificial climate control in a space depends on its function, applying a principle of "heat the people, not the space". A system of five separate climate zones ensure that physical adaptations are minimized, making it possible to transform the LocHal building into one large usable volume.

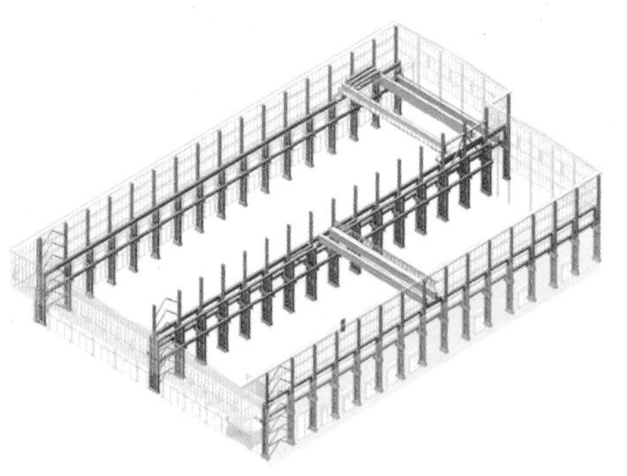

机车大厅 locomotive hall

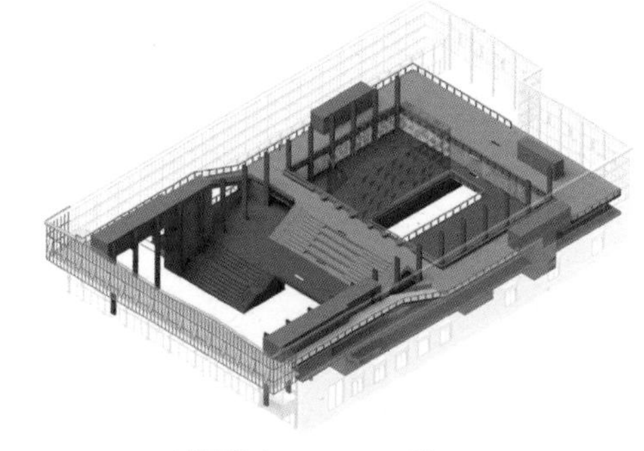

交织建筑 interwoven architecture

可移动挡帘 movable textiles

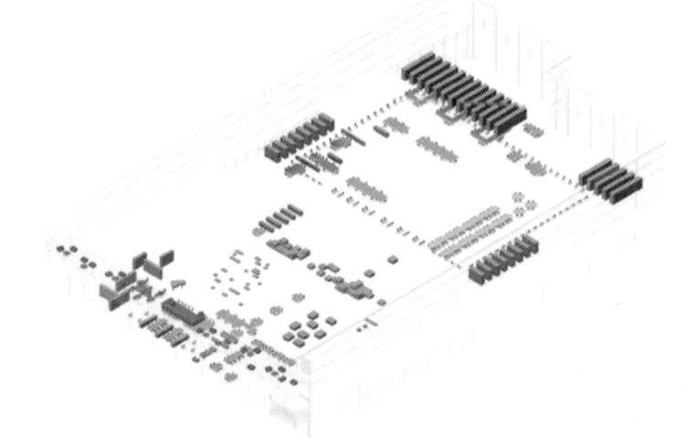

多彩生活 colorful life

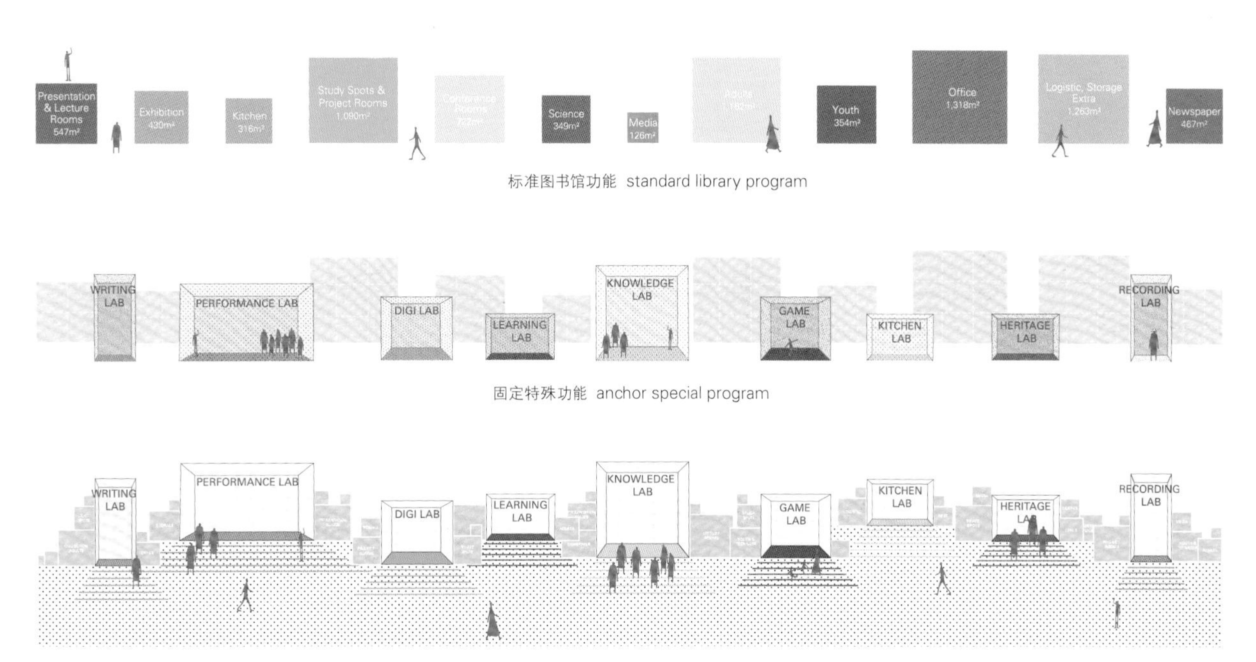

标准图书馆功能 standard library program

固定特殊功能 anchor special program

公共区域 public areas

KennisMakerij

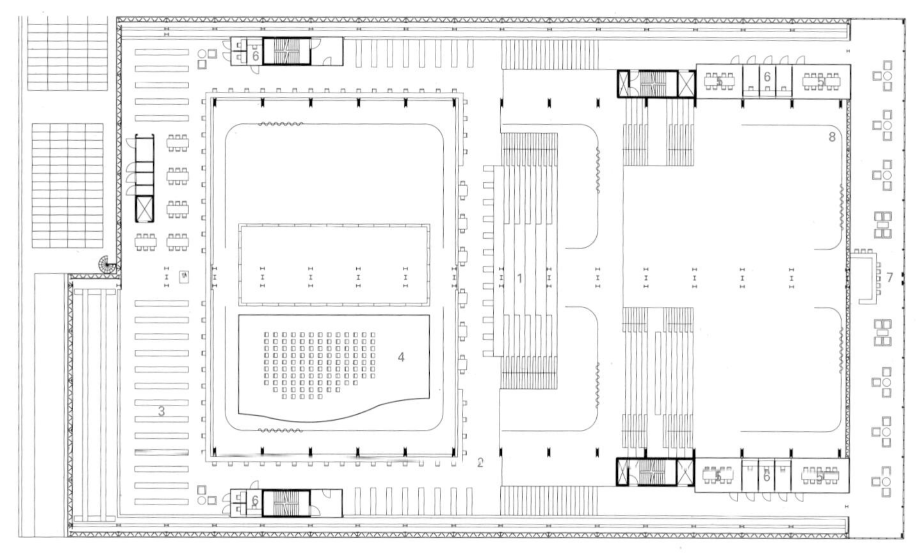

三层和四层 second and third floor

1. 公共楼梯/非正式工作空间
2. 工作空间
3. 科学藏书室
4. 音乐厅
5. 放映室
6. 单独学习室
7. 餐厅休息室
8. 挡帘

1. public staircase, informal working
2. workplaces
3. collection of science
4. concert hall
5. project space
6. concentration space
7. restaurant lounge
8. textile screens

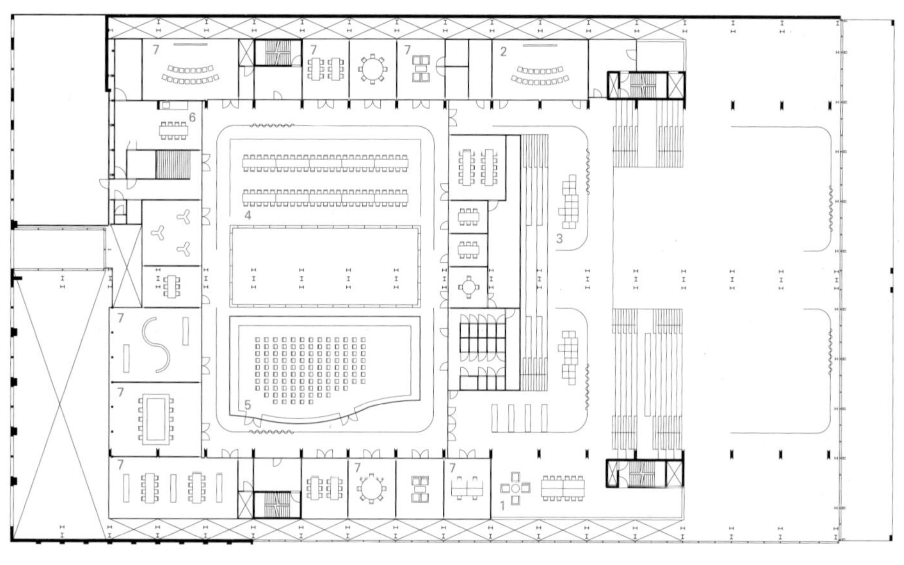

二层 first floor

1. 传统实验室
2. 实验结果讨论室
3. 实验介绍室
4. 工作空间
5. 音乐厅
6. 厨房
7. 会议室

1. heritage lab
2. lab opinion & debate
3. lab introductory
4. workplace
5. concert hall
6. kitchen
7. conference room

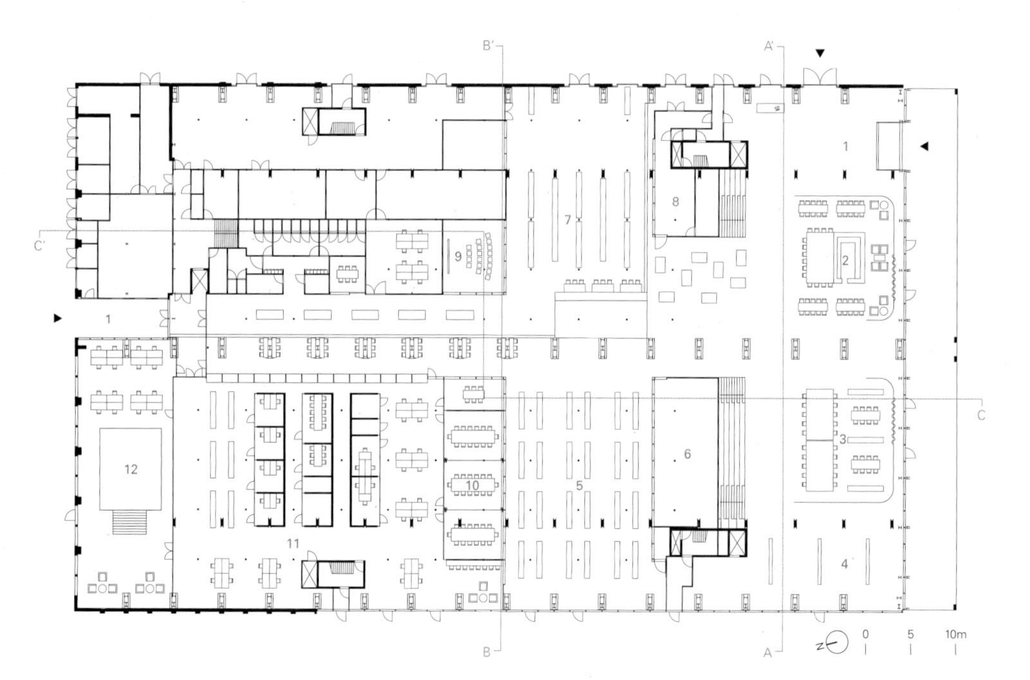

1. 入口
2. 城市咖啡厅
3. 阅览室&开放式讲台
4. 艺术机构展览空间
5. 青少年图书馆
6. 数字实验室
7. 图书馆
8. 烹饪实验室
9. 真人图书馆
10. 会议室
11. 办公室
12. 夹层

1. entrance
2. city cafe
3. reading places & open podium
4. exhibition space art institutions
5. youth library
6. digital lab
7. library
8. cooking lab
9. living library
10. meeting room
11. office
12. mezzanine

一层 ground floor

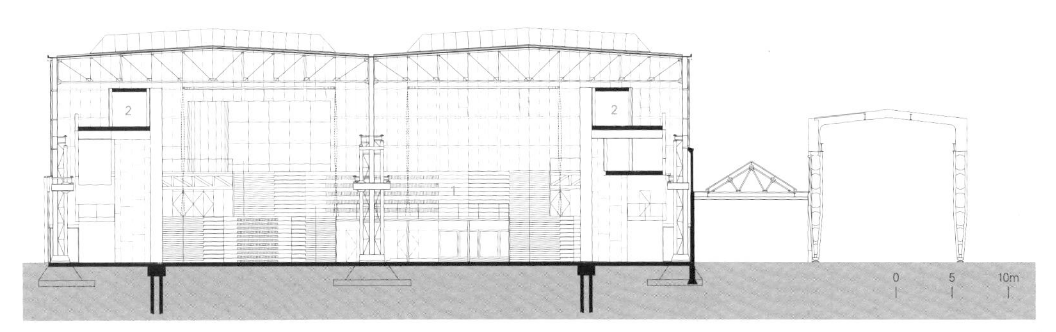

A-A' 剖面图 section A-A'

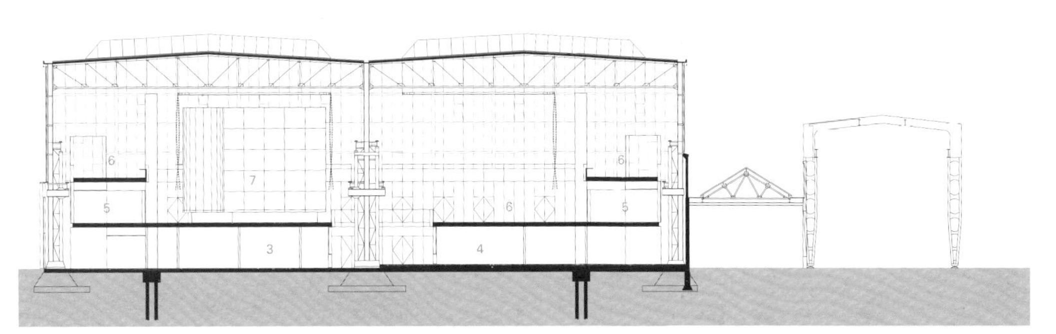

B-B' 剖面图 section B-B'

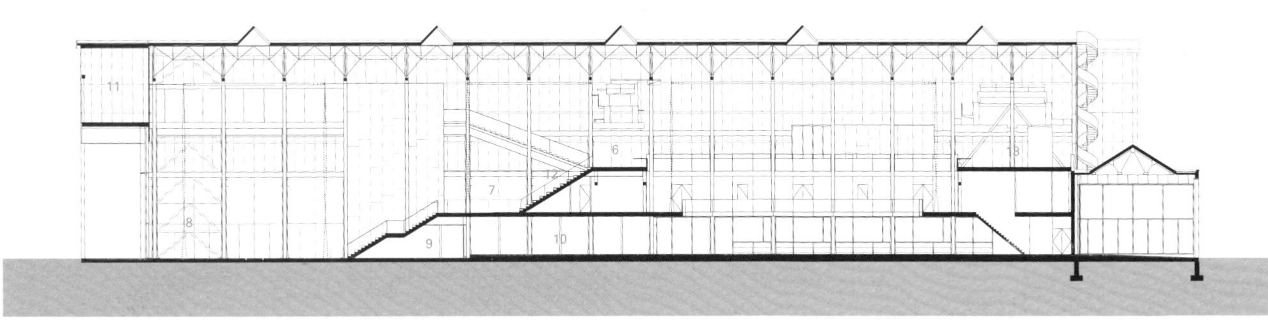

C-C' 剖面图 section C-C'

1. 实验介绍室 2. 放映室 3. 会议室 4. 真人图书馆 5. 会议室 6. 工作空间 7. 音乐厅 8. 阅览室&开放式讲台 9. 数字实验室 10. 青少年图书馆 11. 餐厅休息室 12. 公共楼梯、非正式工作空间 13. 科学藏书室

1. lab introductory 2. project space 3. meeting room 4. living library 5. conference room 6. workplaces 7. concert hall 8. reading place & open podium 9. digital lab 10. youth library 11. restaurant lounge 12. public staircase, informal working 13. collection of science

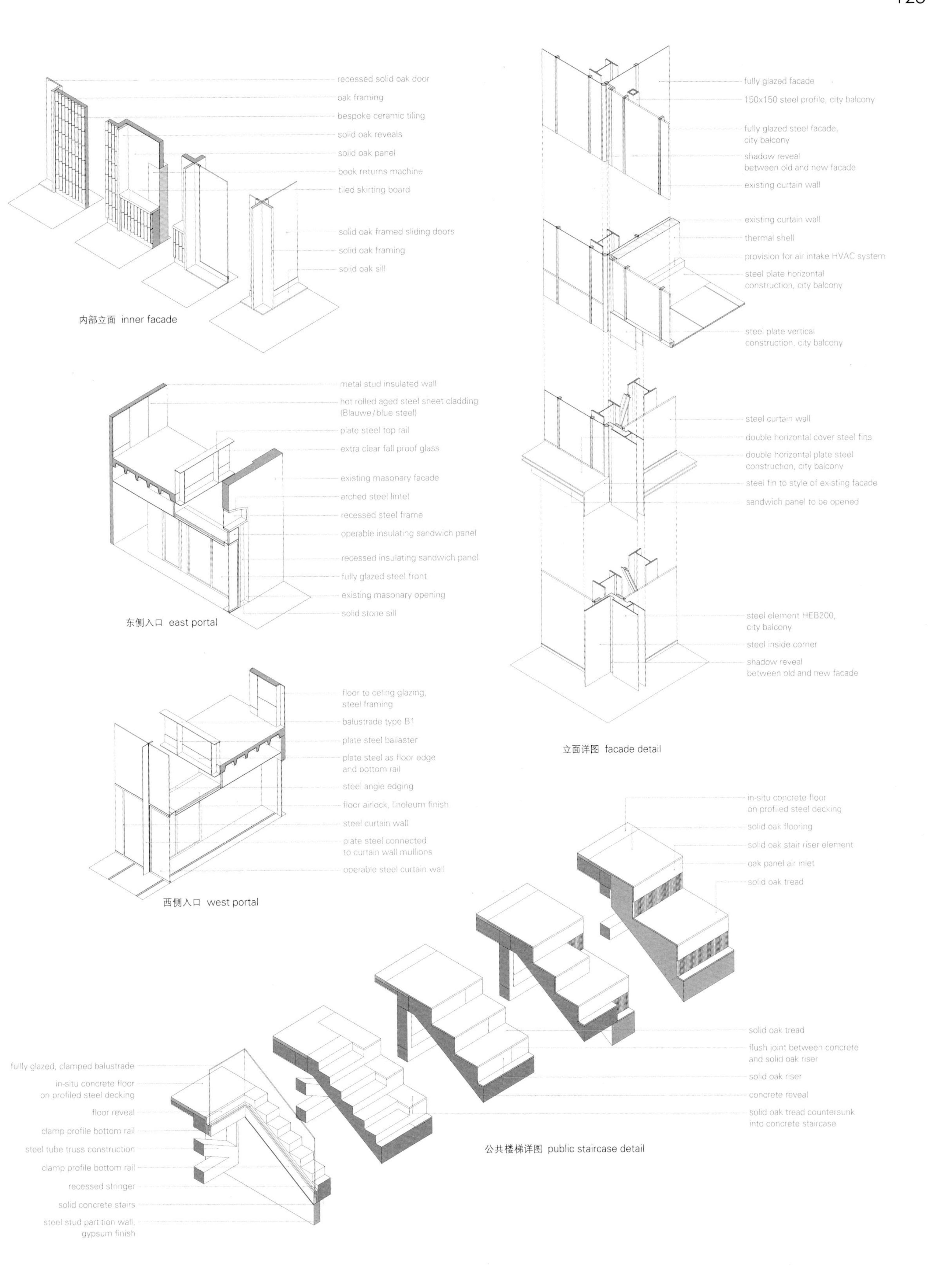
recessed solid oak door
oak framing
bespoke ceramic tiling
solid oak reveals
solid oak panel
book returns machine
tiled skirting board
solid oak framed sliding doors
solid oak framing
solid oak sill
内部立面 inner facade
metal stud insulated wall
hot rolled aged steel sheet cladding (Blauwe/blue steel)
plate steel top rail
extra clear fall proof glass
existing masonary facade
arched steel lintel
recessed steel frame
operable insulating sandwich panel
recessed insulating sandwich panel
fully glazed steel front
existing masonary opening
solid stone sill
东侧入口 east portal
floor to celing glazing, steel framing
balustrade type B1
plate steel ballaster
plate steel as floor edge and bottom rail
steel angle edging
floor airlock, linoleum finish
steel curtain wall
plate steel connected to curtain wall mullions
operable steel curtain wall
西侧入口 west portal
fully glazed facade
150x150 steel profile, city balcony
fully glazed steel facade, city balcony
shadow reveal between old and new facade
existing curtain wall
existing curtain wall
thermal shell
provision for air intake HVAC system
steel plate horizontal construction, city balcony
steel plate vertical construction, city balcony
steel curtain wall
double horizontal cover steel fins
double horizontal plate steel construction, city balcony
steel fin to style of existing facade
sandwich panel to be opened
steel element HEB200, city balcony
steel inside corner
shadow reveal between old and new facade
立面详图 facade detail
in-situ concrete floor on profiled steel decking
solid oak flooring
solid oak stair riser element
oak panel air inlet
solid oak tread
solid oak tread
flush joint between concrete and solid oak riser
solid oak riser
concrete reveal
solid oak tread countersunk into concrete staircase
fullly glazed, clamped balustrade
in-situ concrete floor on profiled steel decking
floor reveal
clamp profile bottom rail
steel tube truss construction
clamp profile bottom rail
recessed stringer
solid concrete stairs
steel stud partition wall, gypsum finish
公共楼梯详图 public staircase detail

项目名称：LocHal / 地点：Tilburg, The Netherlands / 主持建筑师：Civic Architects / 改造：Braaksma & Roos Architectenbureau / 室内与景观设计、挡帘：Inside Outside / Petra Blaisse / 景观设计：Donkergroen / 图书馆&办公室的室内设计：Mecanoo / 施工管理：Stevens van Dijck Bouwmanagers & Adviseurs / 技术顾问（结构设计与工程、装置技术、可持续性、照明、消防、声效设计）：Arup / 顾问：site architect – VDNDP Structural Engineers; structural engineering – F. Wiggers Ingenieursbureau; building physics, acoustics, fire safety – ABT Wassenaar; commercial management – SOM; BREEAM certification – Linneman / 玻璃大厅：Zaanen Spanjers Architecten, Octatube
客户：City of Tilburg / 功能：Bibliotheek Midden-Brabant (public library), Seats2Meet, BKKC, Kunstbalie, catering facilities, events areas / 总建筑面积：11,200m²
开放时间：2019.1 / 摄影师：© Stijn Bollaert (courtesy of the architect)

2.4 Films
2.4 Films

卢森堡大学学习中心
Luxembourg Learning Center

Valentiny hvp Architects

新建的大学图书馆中心将历史悠久的锅炉厂融入了现代建筑设计之中

New university library center integrates historic blast furnaces into contemporary architectural design

都市概念

“高炉排楼”是卢森堡一个独具创意的城市规划项目。图书馆位于项目中心地带，其建筑风格展示了该地区的工业遗迹。

该项目保留了现有的结构要素、A号和B号高炉及其附属建筑。新建成的中心建筑用于教学和研究，它就是围绕这些历史特色而建造的，并且创造了一个集工作、文化、休闲、居住为一体的多功能城市校园，从而创造了一个引人注目的“生活空间”。

工业遗产

“Möllerei”（用于储存焦炭和铁矿石等原材料）的整个外部体量都得到了保留，南部有8个分隔间用于生产铁；北部17个分隔间全部清空，只留下支撑钢结构和一个矿仓。现在这里变成了大学图书馆，也就是卢森堡大学学习中心。

项目实施

图书馆位于场地的南部，形成了社会文化中心；教学中心位于场地北部，由知识之家确定边界。

图书馆的一边是学院广场，另一边是高炉广场。步行空间连接着一连串的广场。共同构成了一个同时具有历史特征和当代特征的空间整体。

功能概念

该项目包括四大功能：入口大厅和会议室、咨询和工作空间、行政和自助餐厅。

用户的主要区域包括藏品和所有多媒体工具，都位于将以前的“Möllerei”翻新后的大空间内。文件档案存放在地下室中。

入口大厅和独立会议室以及图书馆管理处都被安置在新建体量中。第一个体量是椭圆形的，人们可以通过高炉广场进入。而第二个体量由三个矩形体量构成，位于较低的一层，可以俯瞰学术广场。

建筑概念

该项目的主要目标之一是创造一个开放而受欢迎的学习空间。

公众可以从一个超过100m长的大型平台进入图书馆，这个平台像飞碟一样围绕着矿仓周围的五层楼，老建筑被清理、加固和重新粉刷，保留了建筑过去活动的所有痕迹。

一个新的外围护结构将这些空间覆盖，与屋顶形成了一个风格统一且有特色的整体。该外围护结构的设计与内部功能的分布密切相关，北面和东面的外墙板，也就是阅读区的位置，是平面的。在西侧，凸出的六边形元素能够调节光线，营造出自然、间接而又柔和的效果。在六个金字塔形立面中，朝北的两个立面是透明的，而其他的立面则是不透明的。

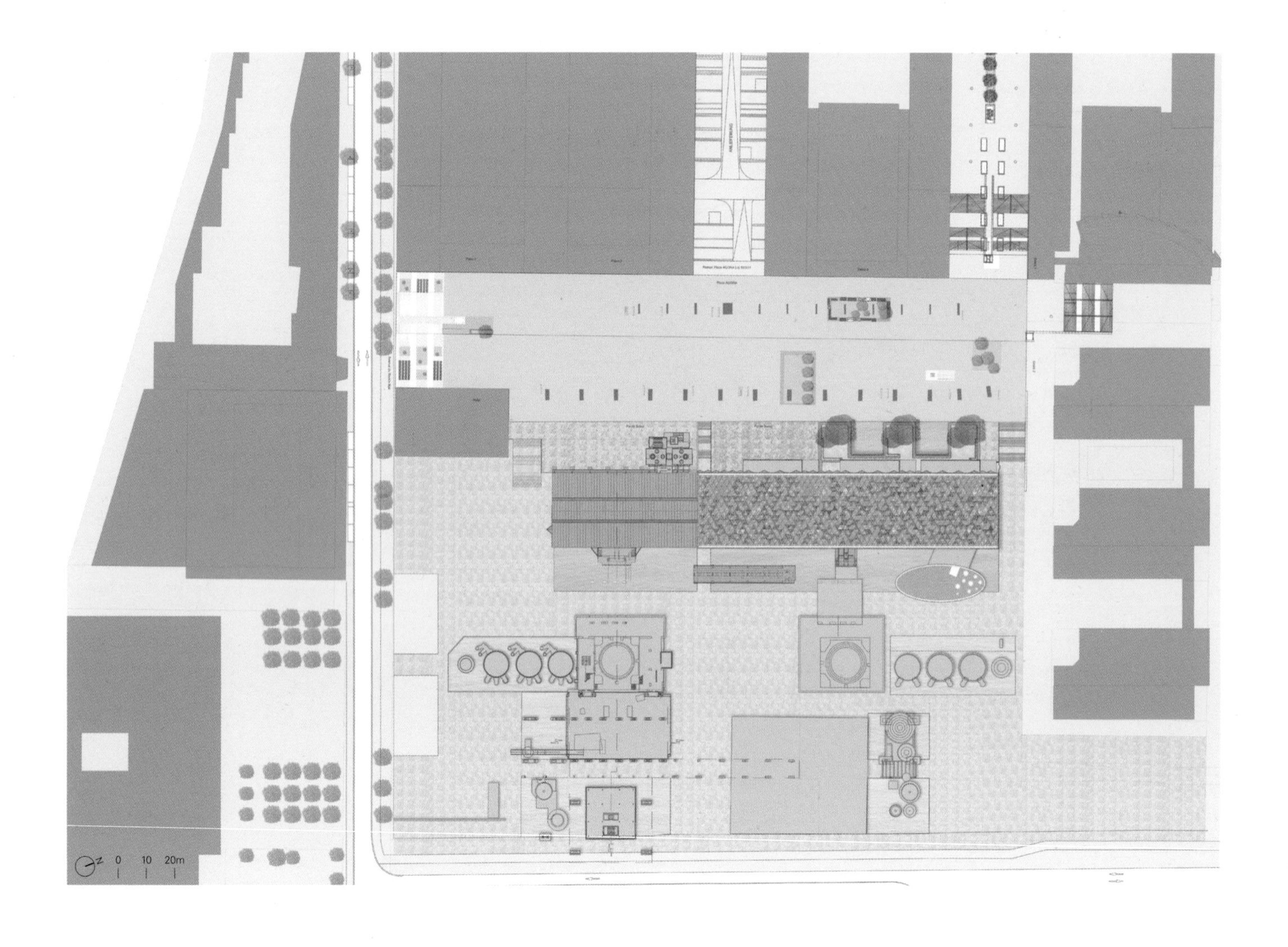

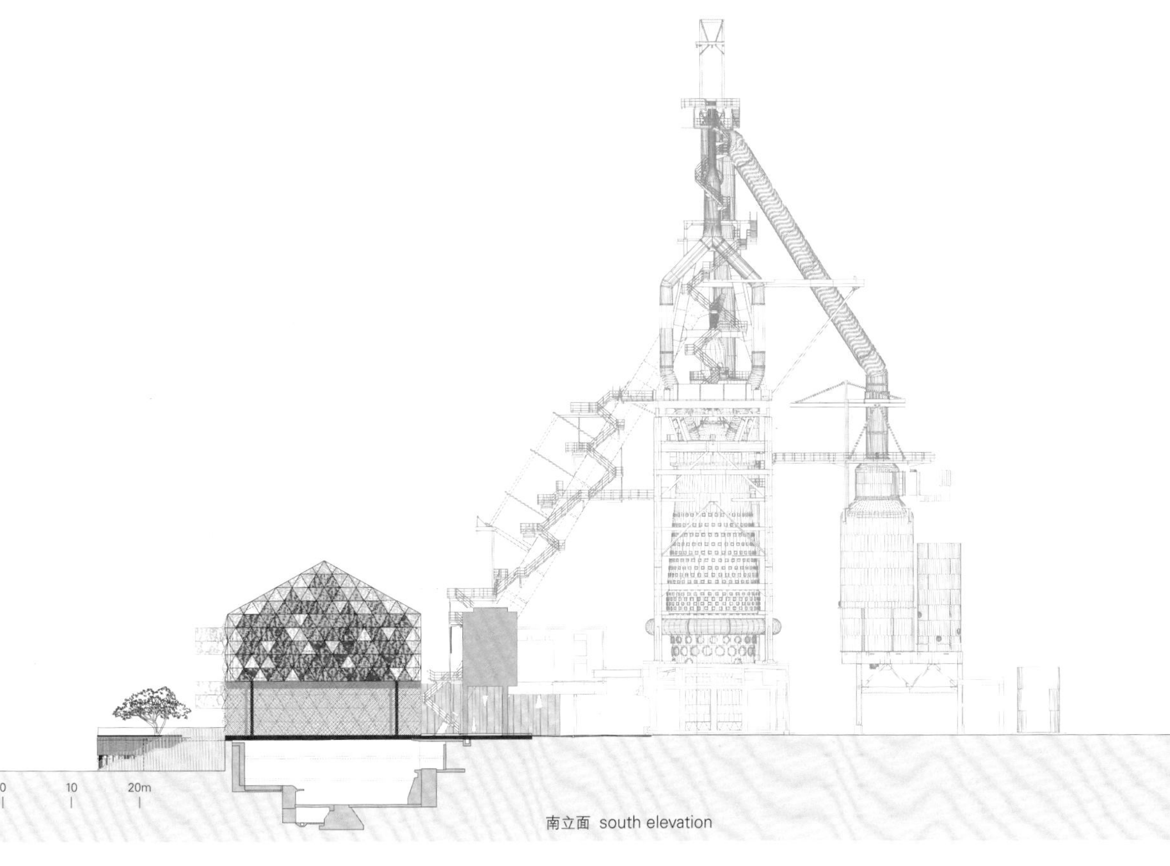

南立面 south elevation

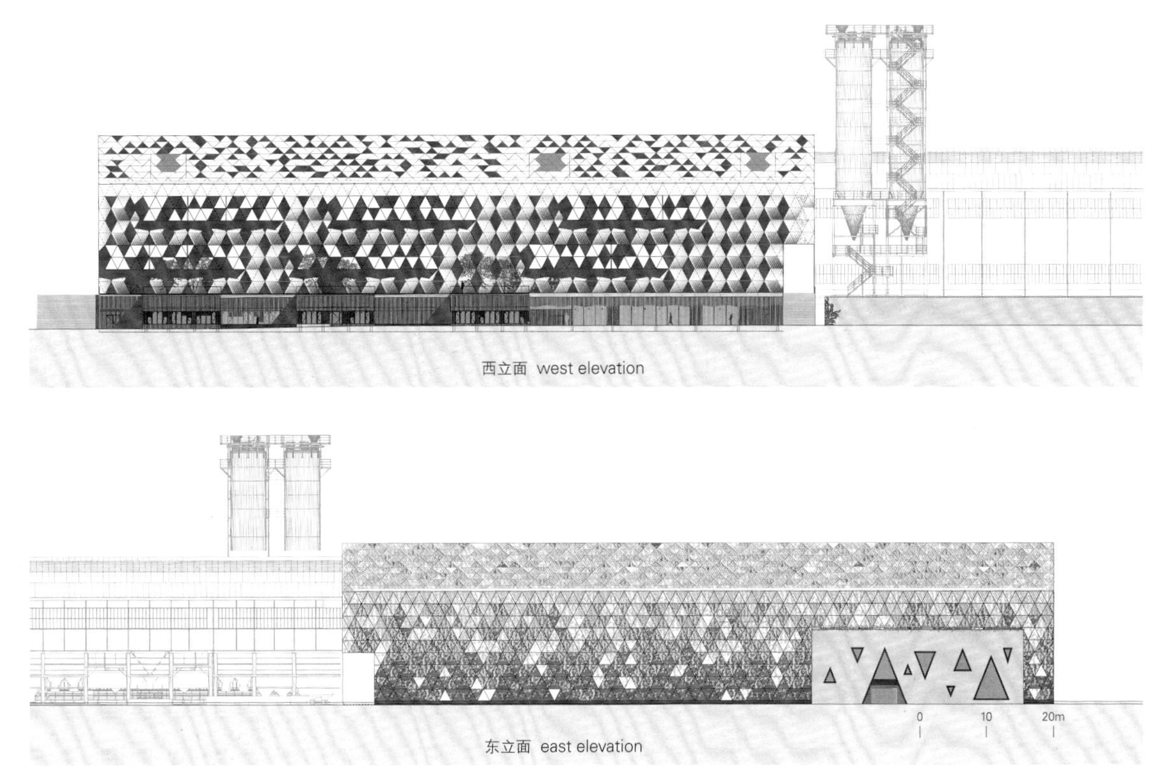

西立面 west elevation

东立面 east elevation

可改变亮度的丝网印刷玻璃起到了遮阳效果，减少了太阳得热和眩光，同时使建筑外观呈现出了大理石的效果。从内部看，外立面更加通透，呈现出开放和磨砂的三角形表面。

椭圆形入口体量被白色板条覆盖，这与建筑历史印迹产生了鲜明的对比。

室内空间的特点是具有流动性和通透感，同时还兼顾了遗留材料的状态与工业属性。无论是垂直维度上的混凝土墙，还是涂有白漆的墙面，都覆盖有相应的吸声薄板，从而和烟灰色的地面和吊顶保持协调一致。

Urban Concept

The district of Terrasse des Hauts Fourneaux (Terrace of Blast Furnaces) is a unique and innovative urban planning project in Luxembourg. At its heart is the Maison du Livre, whose architecture showcases the industrial heritage of the area. The project retains elements of the existing structure, Blast Furnaces A and B, and their ancillary buildings. The new center, dedicated to teaching and research, is structured around these historic features, and creates a multifunctional urban campus where work, culture, leisure and housing meet to create an attractive "living space".

Industrial Heritage

The entire external volume of the "Möllerei" – which had been used to store raw materials coke and iron ore – was preserved, as have eight of the south bays, where pig iron was produced. Seventeen bays in the northern part were emptied, leaving only the supporting steel structure and an ore silo. This now houses the University Library – the Luxembourg Learning Center.

项目名称：Luxembourg Learning Center
地点：Esch-sur-Alzette, Luxembourg
建筑师：Valentiny HVP Architects
合作方：RMC Consulting s.à.r.l.
工程师：Bollinger + Grohmann Ingenieure
总建筑面积：15,000mm²
总建筑体量：84,100m²
最终预算：59,500,000 euros含税
竣工时间：2018
摄影师：
©Michel Zavagno / Blitz Agency (courtesy of the architect) - p.130, p.132, p.133 bottom-right, p.136, p.140, p.141
©Guy Jallay (courtesy of the architect) - p.126~127, p.131, p.137
Courtesy of the architect - p.129

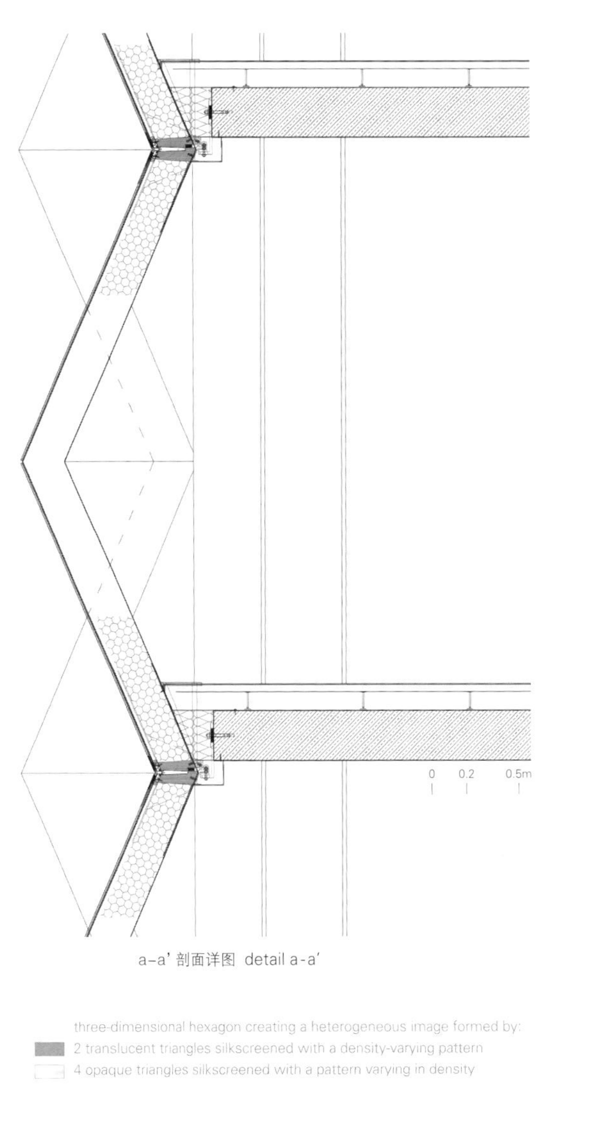

a-a' 剖面详图 detail a-a'

three-dimensional hexagon creating a heterogeneous image formed by:

2 translucent triangles silkscreened with a density-varying pattern
4 opaque triangles silkscreened with a pattern varying in density

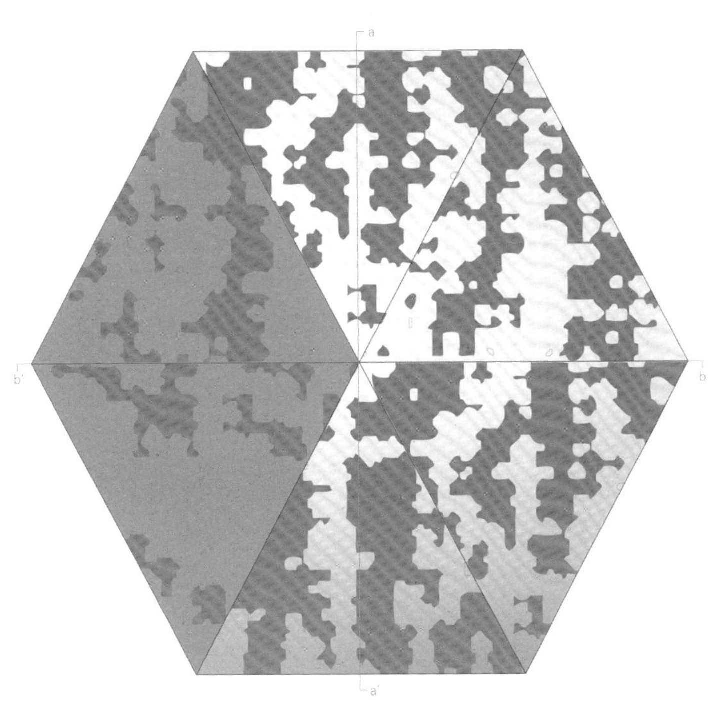

详图1——主立面构成
detail 1 _ main facade composition

b-b' 剖面详图 detail b-b'

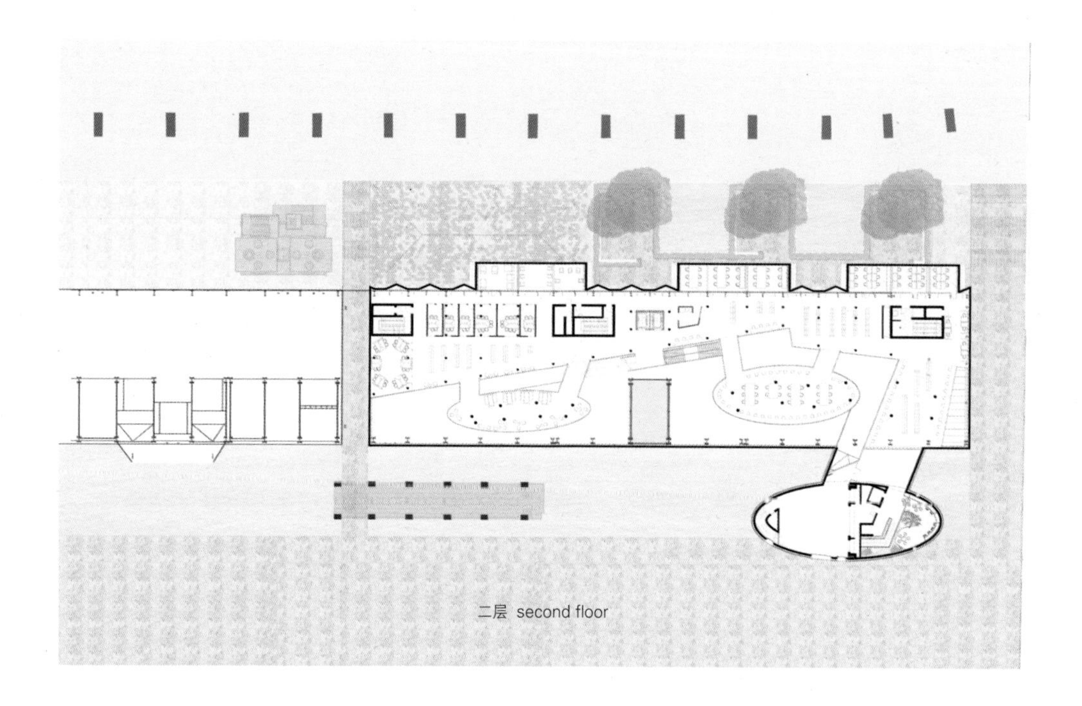

二层 second floor

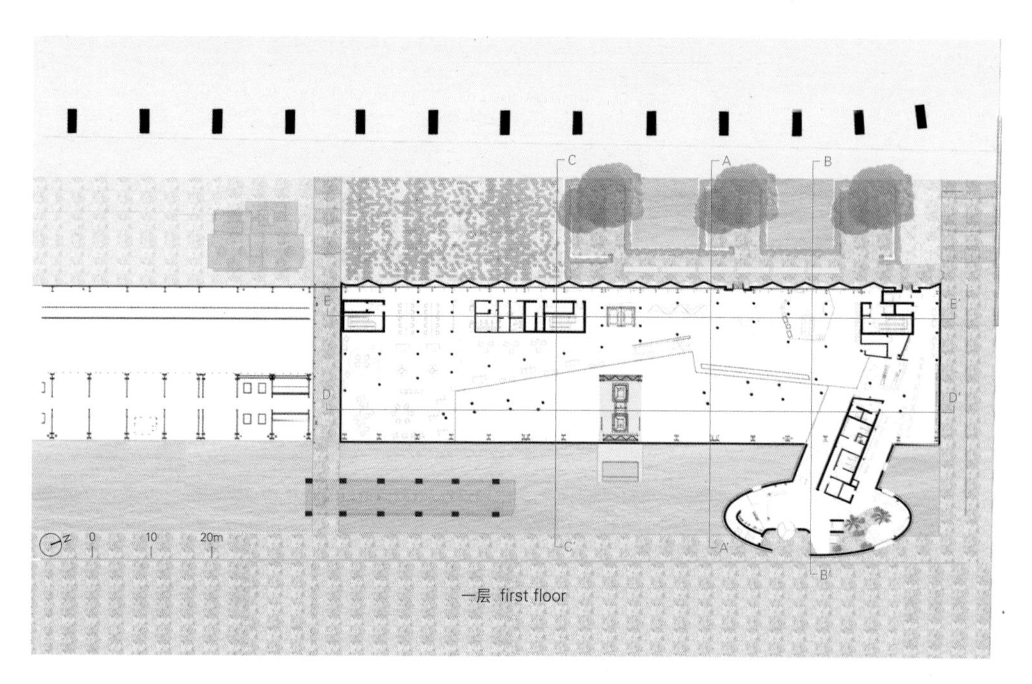

一层 first floor

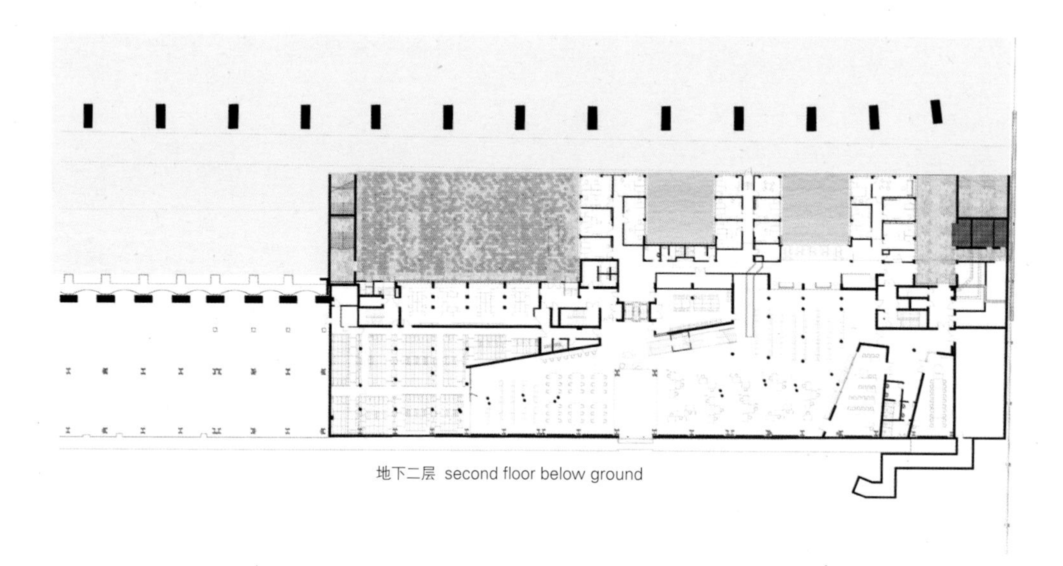

地下二层 second floor below ground

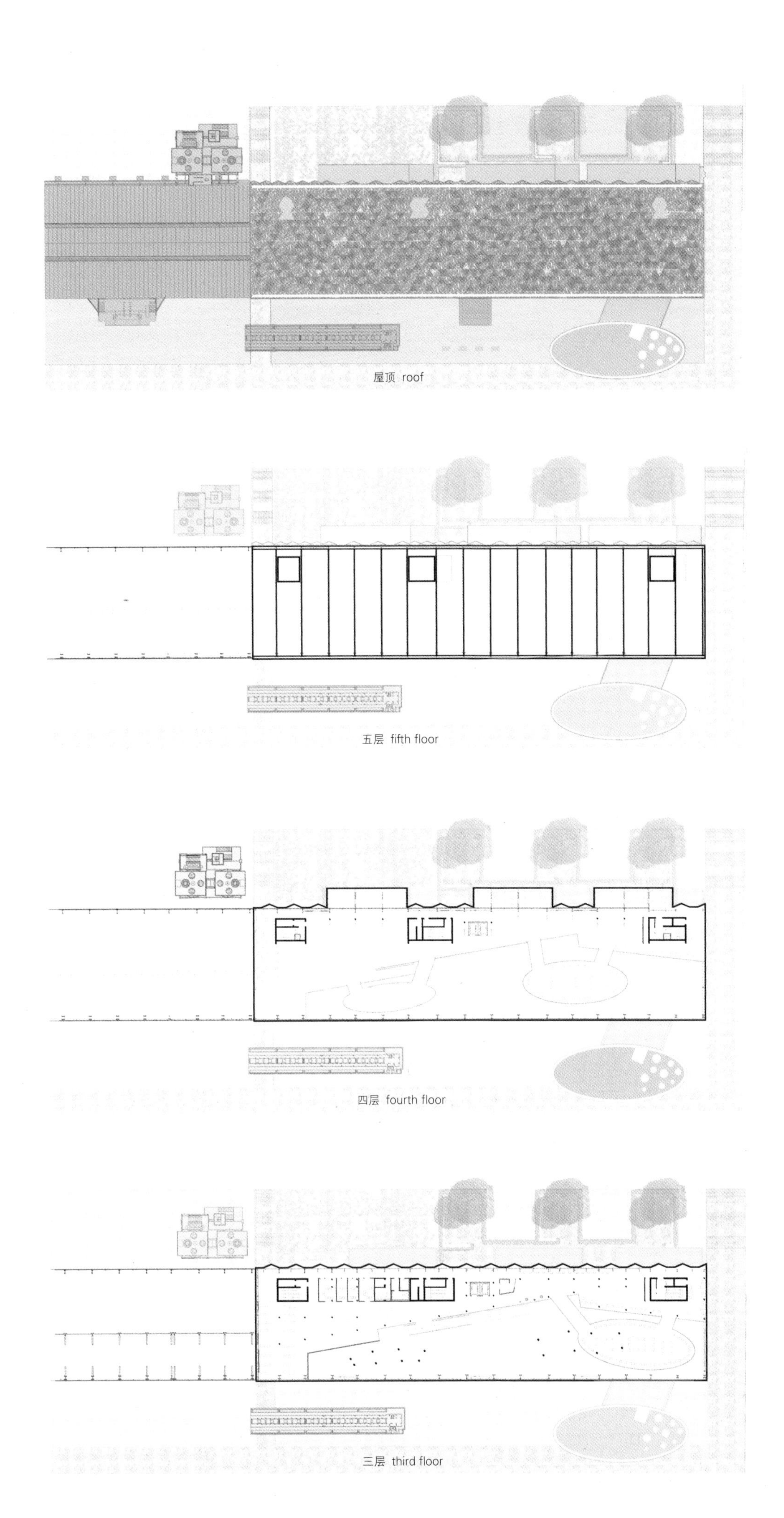

屋顶 roof

五层 fifth floor

四层 fourth floor

三层 third floor

Implementation

The Maison du Livre (House of the Book) is located in the southern part of the site, forming the socio-cultural center. The teaching center, defined by the Maison du Savoir (House of Knowledge), is located to the north of the site.

The Maison du Livre is surrounded by the Place de Académie on one side and on the other by the Place des Hauts Fourneaux, lively pedestrian spaces articulated as a succession of squares. Together these form a simultaneously historic and contemporary spatial ensemble with a strong identity.

Functional Concept

The program consists of four main functions: entrance hall and conference room, consultation and working spaces, administration and cafeteria.

The main areas for users, including the collections and all multimedia tools, are housed in the large refurbished volume of the former "Möllerei". In the basements are the document archives. The entrance hall and independent conference room, as well as the library administration, are housed in new volumes. The first, elliptical in shape, allows public access through the Place des Hauts Fourneaux. The second, in the form of three rectangular volumes, is located on the lower level overlooking the Place de l'Académie.

Architectural Concept

One of the major project objectives was to create an open and welcoming learning space.

Large platforms, over 100m long and accessible to the public, appear on five levels, like flying saucers around the ore silo. The old structures have been cleaned, reinforced and repainted, preserving all traces of the building's past activity.

A new envelope covers these, forming with the roof a homogeneous and characterful ensemble. The design of this envelope is closely linked to the distribution of interior functions: the north and east facade panels, where the reading areas are located, are flat. On the west side, projecting hexagonal elements modulate light to create natural, indirect and soft effects. Two of the six pyramid facets, those facing north, are transparent, whereas the others are opaque.

Variable intensity screen-printed glass provides solar protection, reducing heat gain and glare, whilst giving the building exterior a marbled effect. From the inside, the façade is more transparent, with a play of open and opaque triangular surfaces.

The elliptical entrance volume is covered by white slats, clearly distinguishable from the envelope of the historic part of the project.

The interior spaces are characterized by fluidity and luminosity, but also by the sobriety of the materials and their industrial character. The vertical surfaces, either in raw concrete or painted white, are next to the anthracite grey carpet floors and the ceilings streaked with acoustic lamellae.

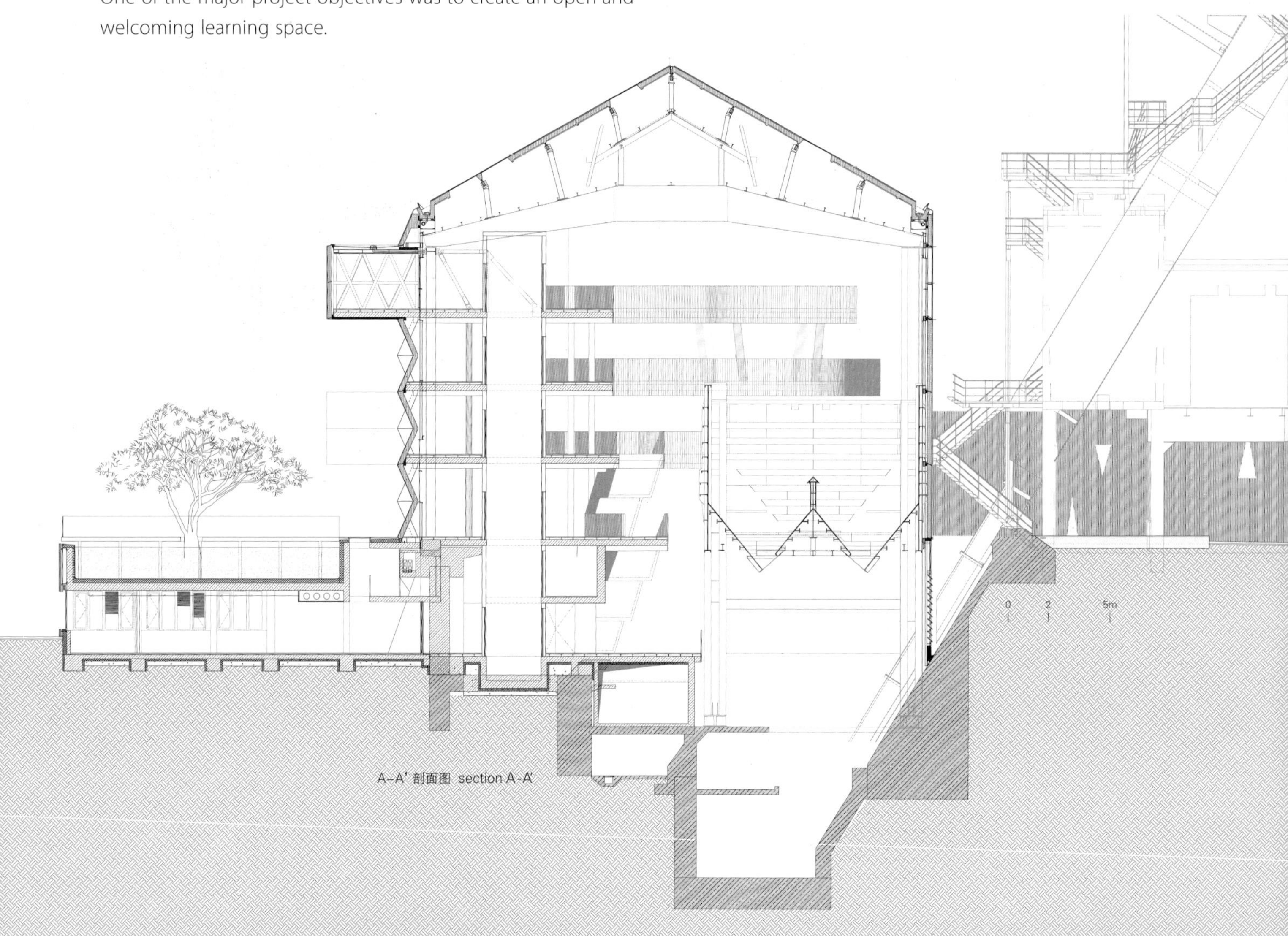

A–A' 剖面图 section A-A'

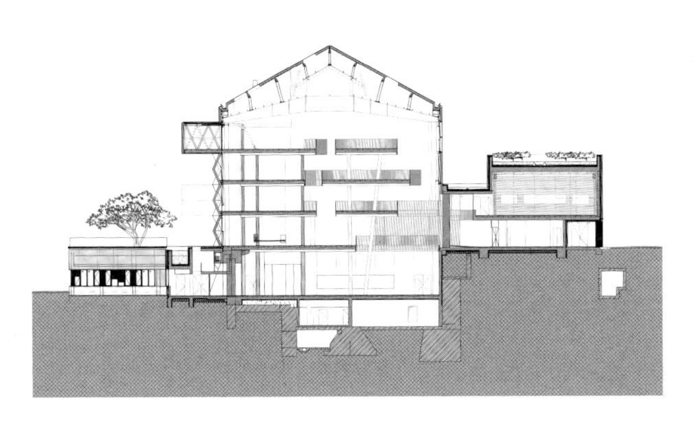

B-B' 剖面图 section B-B′

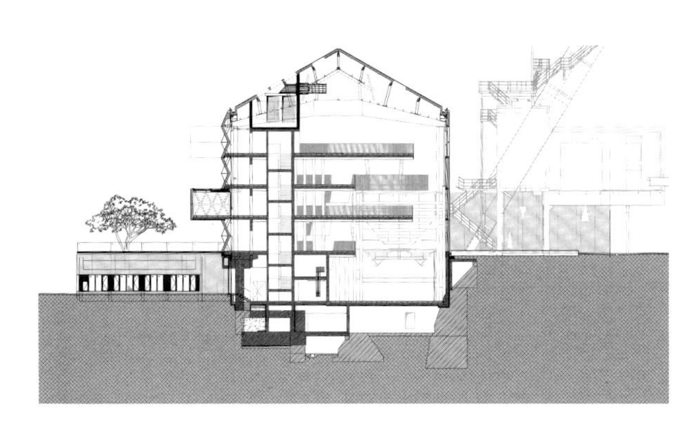

C-C' 剖面图 section C-C′

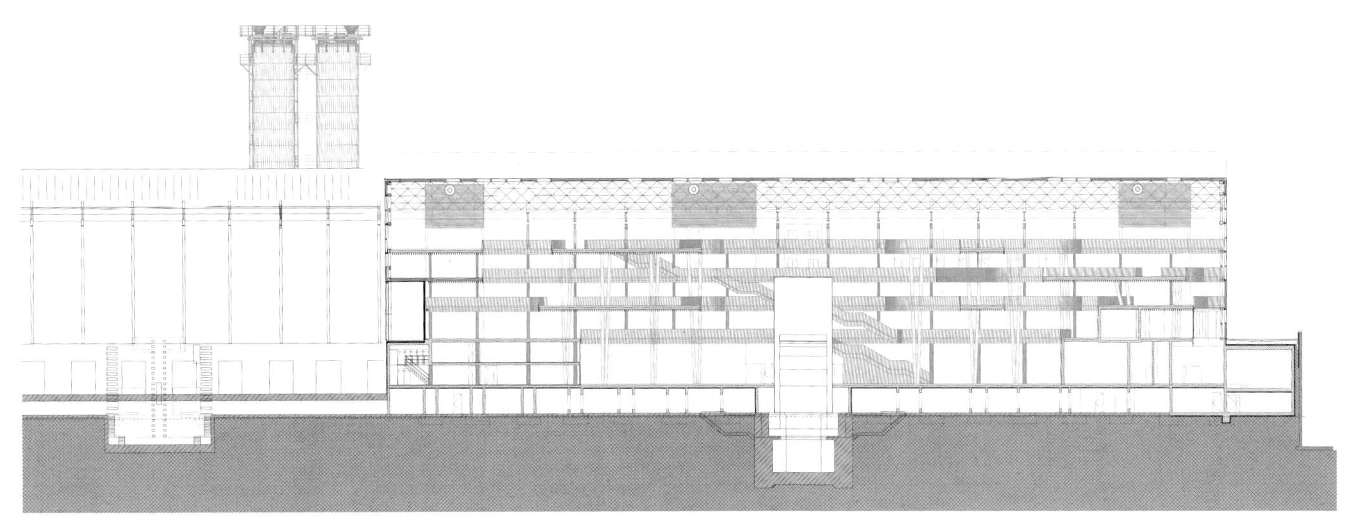

D-D' 剖面图 section D-D′

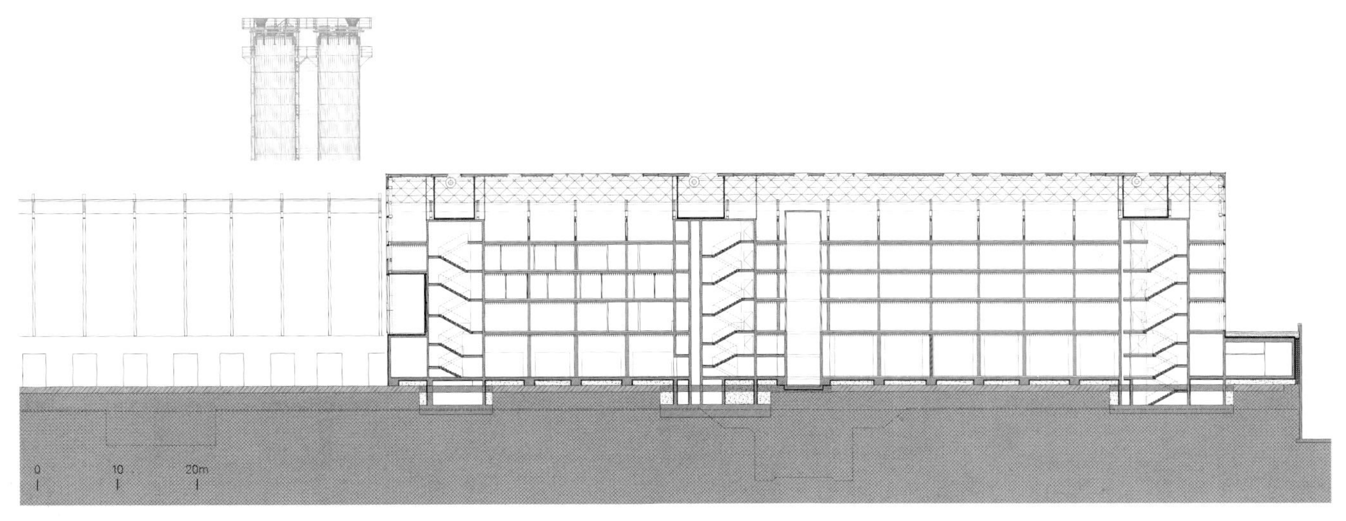

E-E' 剖面图 section E-E′

BBDO

布鲁塞尔 BBDO 广告公司
Advertising Agency BBDO Brussels

ZAmpone Architectuur

赞普设计为比利时首都提供灵活的新 BBDO 办公空间

ZAmpone's intervention provides flexible new BBDO office space in the Belgian capital

创意广告公司BBDO在布鲁塞尔市莫伦贝克–圣让镇拥有一栋大型受保护建筑。1999年，相关单位对大楼后面的仓库进行了第一次翻修。由于公司内部进行了多次重组，所以BBDO委托公司为其办公室开发了一个在概念上富有创意的空间。

从经济可行的角度来说，空间得到有效利用和具有更高的容量是这个项目的重要条件。赞普公司的设计通过强调三个原则来赞美开放空间：空间活力、声学设计以及对自然光的利用。

在“新的工作方式”中，移动并适应环境的不是空间，而是人，人是静态建筑中的动态用户。赞普公司的设计目的不仅是打造一个活力十足的空间，同时也提出了一种自我维持的财务模式来支持这种新的工作理念：BBDO清空多余面积，把类似接待处和自助餐厅的空间纳入公共使用中。这一设计同样也是在突出“空间活力”这一要素。

圈定噪声区域是设计中的第二大要素。在开放的工作空间中，通过将会议限制在规定单元内，使音量得到有效的控制。非正式会议和谈话可以在餐厅、楼廊内部和交通流线周围进行。

自然光的利用是第三个要素。众所周知，员工在有自然光的空间里工作得更好，欣赏美丽的城市风光更加让他们感到心旷神怡。新的工作空间被布置在靠近中庭的窗边，环绕整个中央挑空空间。会议室故意不提供自然光照，以限制占用时间，并且阻止员工长时间占据这里小憩。

为了优化设计，设计师做了一些基本的调整：主入口被移往建筑的一边与停车场相连。这样，建筑巨大的中庭不仅作为一个流通空间，它还能让来访者直接看到公司的“创意之心”。

建筑的正交网格被对角线形式的改造所打破。三角形的空间创造了新颖有趣的视角。这种对角线形式的设计方法主要应用在走廊（供CEO们举办非正式会议、讨论和讲座的地方）和让接待区更具活力的图书馆区域。

建筑的一层容纳了可与建筑内其他公司共享的集体使用功能，包括餐厅、走廊讲坛以及会议室。

在二层，围绕建筑中央挑空区域的一扇窗户可以让每一层都看得见BBDO的“创意之心”，来访者完全被吸引到办公室的内部工作中；这一层的其他服务设施都经过精心安排，以便使彼此之间的互动能够有机地开展。

最具活力的BBDO团队位于三层。这些用户整天穿梭于不同的服务项目和部门之间，所以用户从不长时间占用同一个办公空间。

电源插座与数据点一直都是灵活办公理念的一大挑战，对此，赞普公司设计了桌间移动式钢管构件，将电缆引向正确的位置。同时它们也被当成了各种隔墙和声学材料的媒介。唯一情况就是这些材料不能阻止光线在空间内传播。钢构件还以简单和线件的方式界定了区域，如悬挂的植物、衣帽架或电灯：个性与灵活性到了这里就不再彼此排斥了。

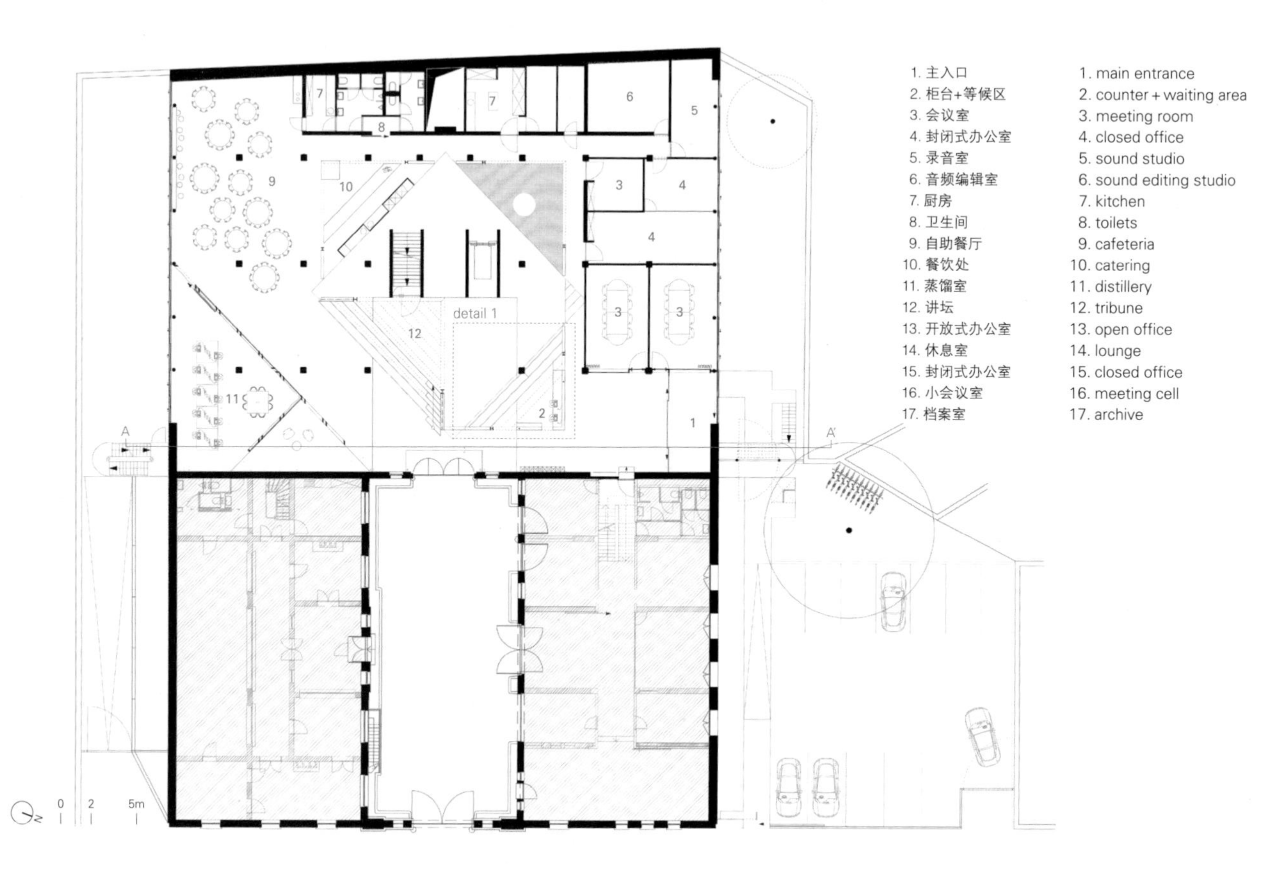

一层 ground floor

Creative agency BBDO occupies a large, protected building in Sint-Jans-Molenbeek, Brussels. The first renovation, of the warehouses at the back of the building, took place in 1999. As a result of a number of internal reorganisations of the company, BBDO commissioned ZAmpone to develop a new spatial concept for their office.
Economic viability, efficient use of space and higher capacity were important conditions of the assignment. ZAmpone's design celebrates open space by placing importance on three principles: dynamics, acoustics and daylight.
In the "new way of working", it is not space but the human being that moves and adapts to a situation – a dynamic user in a static building. ZAmpone's aim was not only to create space that makes dynamism possible, but also to propose a self-sustaining financial model to support this new work ethic. If BBDO frees up surplus square meters, and makes spaces such as reception areas and cafeterias open for common use, that is also "dynamics".
Delineating and defining noisy zones was the second important principle within the design. In the open workspaces, sound is kept under control by allowing meetings only within

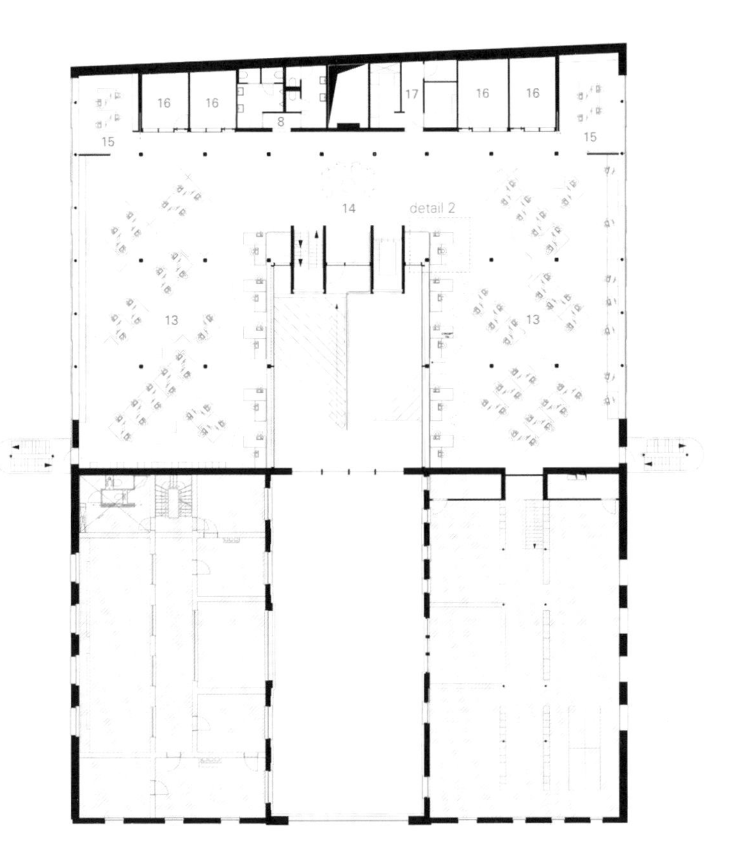

三层 second floor

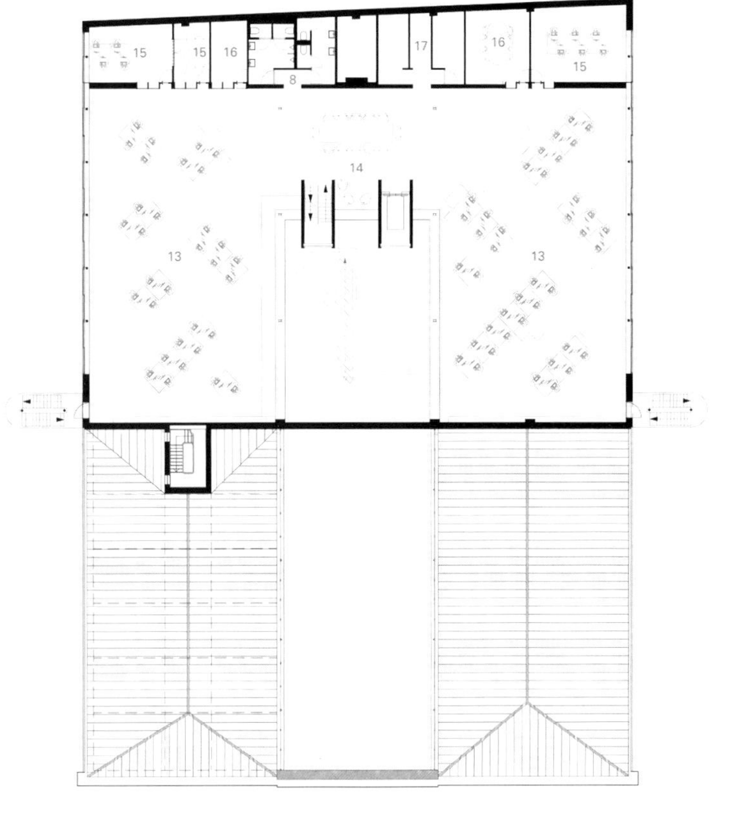

四层 third floor

the provided "cells". Informal meetings and chat take place in the restaurant, in the gallery, and around the circulation routes.

Daylight was the third crucial factor. It is known that employees operate better in spaces with natural light, and a great view over a beautiful urban landscape contributes to their enjoyment. New workspaces are organized close to the windows and around the central void. Meeting rooms are deliberately not provided with daylight, so as to limit their occupation time and discourage employees from long-term "nesting" there.

A basic intervention was necessary to optimize the design: the main entrance was moved to the side of the building, which is connected to the parking lot. This enabled the enormous hall in the middle of the building to become more than just a circulation space. It gives the visitor an immediate insight into the "creative heart" of the company.

Diagonal interventions help break the orthogonal grid. Triangular zones create new, more interesting views. This diagonal approach manifests itself in the gallery (where the CEOs hold informal meetings, discussions, presentations and lectures) and in the library which animates the reception area.

The ground floor accomodates collective functions shared with the other companies in the building, including the restaurant, the gallery and the meeting rooms.

On the first floor, a window around the void at the center of the building allows the "creative heart" of BBDO to be visible from every floor. The visitor is fully drawn into the internal workings of the office; the other services on this floor were carefully organized so as to enable interactions to develop organically.

The most dynamic team of BBDO is housed on the second floor. These users move throughout the day between different services and departments, so maximum occupation of the desk-space never occurs here.

Electrical outlets and data points are always a challenge for a flexible office concept. ZAmpone designed mobile tubular steel elements between the desks that lead cables to the right place. These are also support structures acoustic partitions of various different materials. The only condition for the material choice here was that they must not stop light from penetrating the spaces. Steel elements also define zones in a simple and linear way. A hanging plant, a coat rack, a light: personalization and flexibility are no longer mutually exclusive.

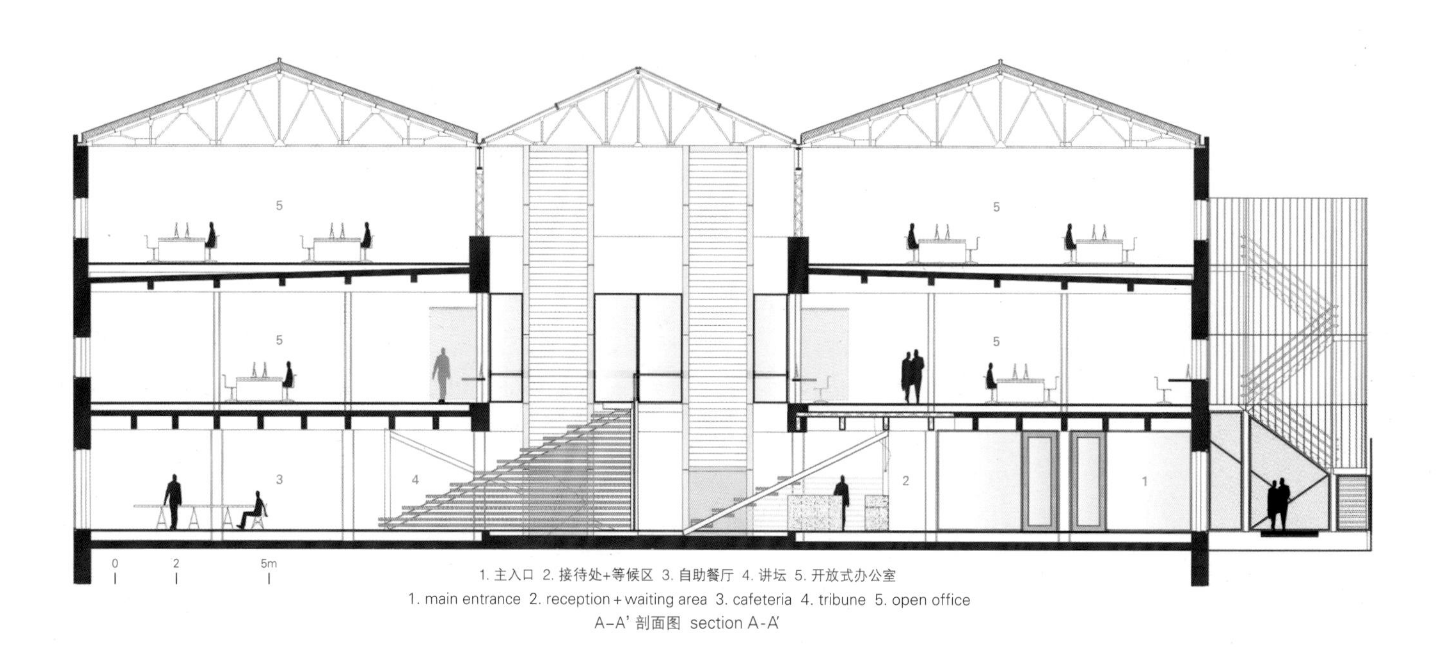

1. 主入口 2. 接待处+等候区 3. 自助餐厅 4. 讲坛 5. 开放式办公室
1. main entrance 2. reception + waiting area 3. cafeteria 4. tribune 5. open office
A-A' 剖面图 section A-A'

项目名称：BBDO Brussels / 地点：Molenbeek-Saint-Jean, Brussels / 建筑师：ZAmpone Architectuur / 主持建筑师：Bart Van Leeuw (partner senior architect) & Kevin Van Steenbergen (projectarchitect) / 结构工程师：UTIL – Pieter Ochelen / 技术工程师：BOTEC nv / 承包商：Detender NV / 总建筑面积：2,695m²
竣工时间：2018 / 摄影师：©Tim Van De Velde (courtesy of the architect)

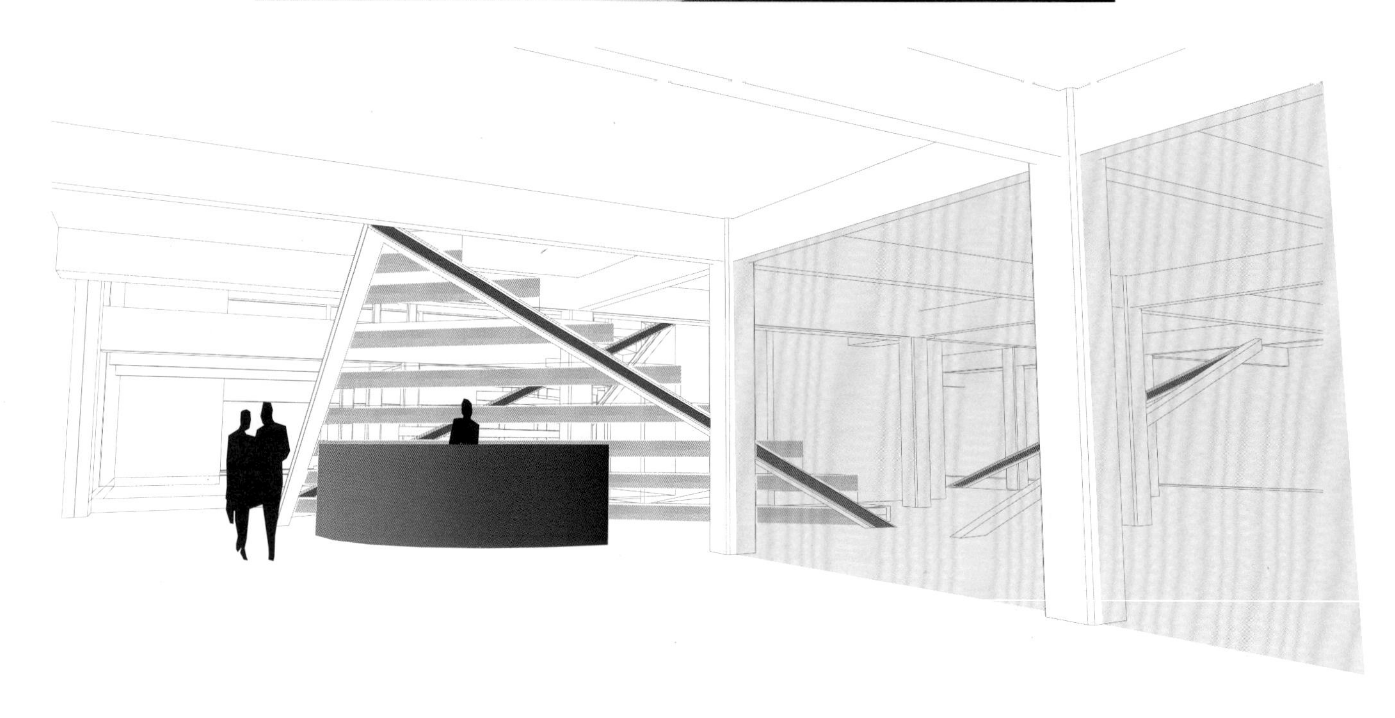

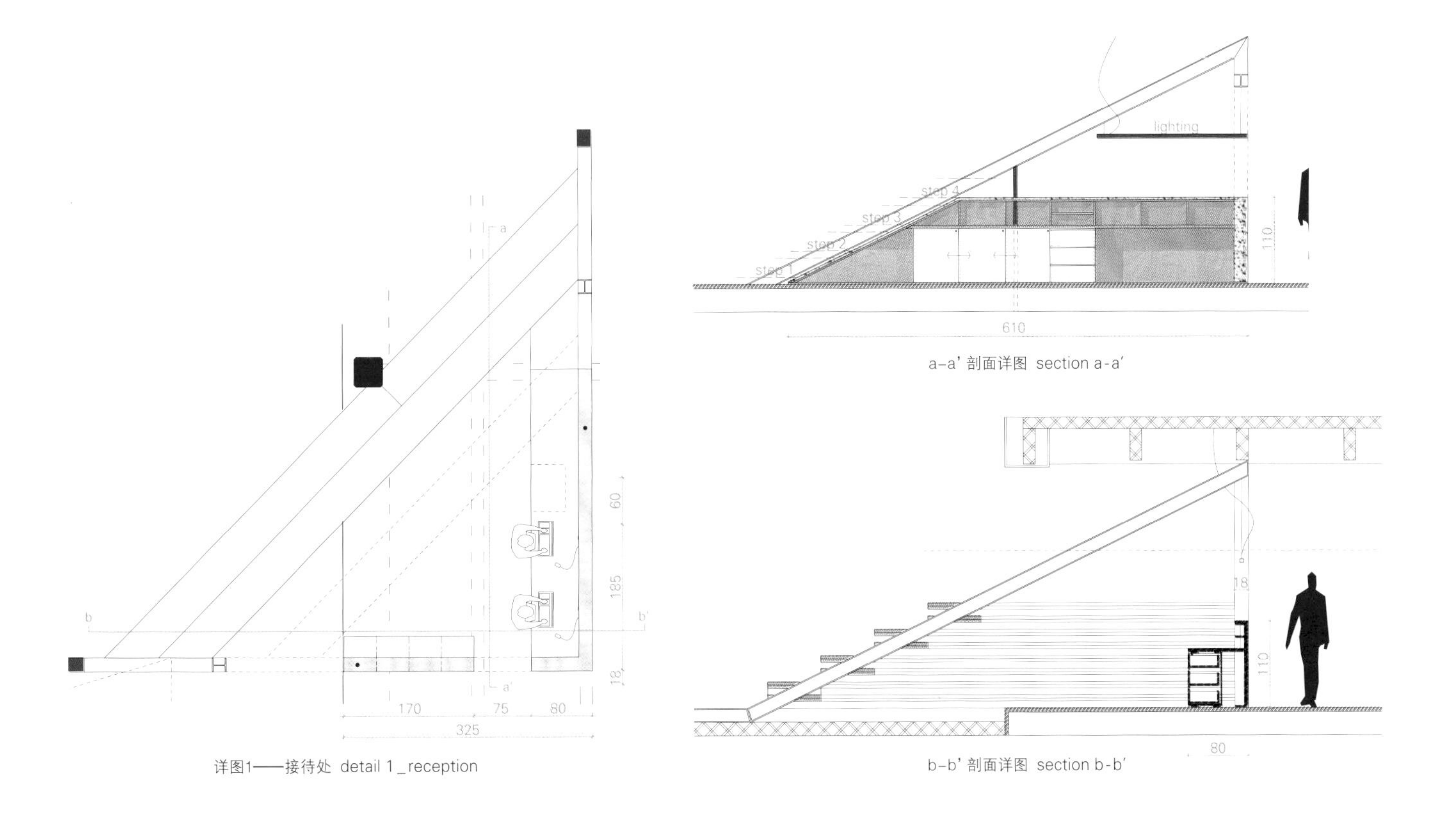

详图1——接待处 detail 1_reception

a-a' 剖面详图 section a-a'

b-b' 剖面详图 section b-b'

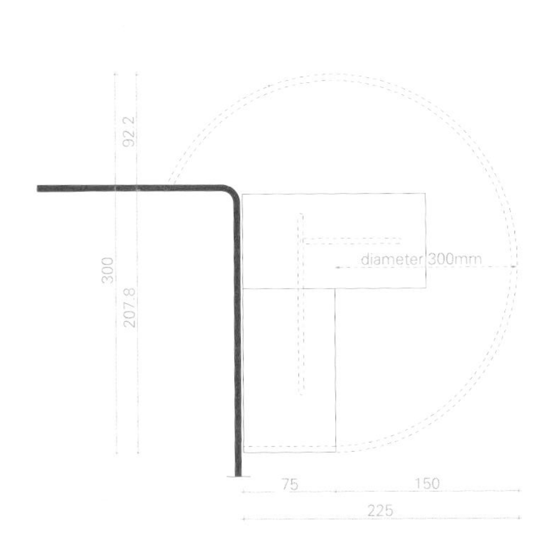

详图2 detail 2

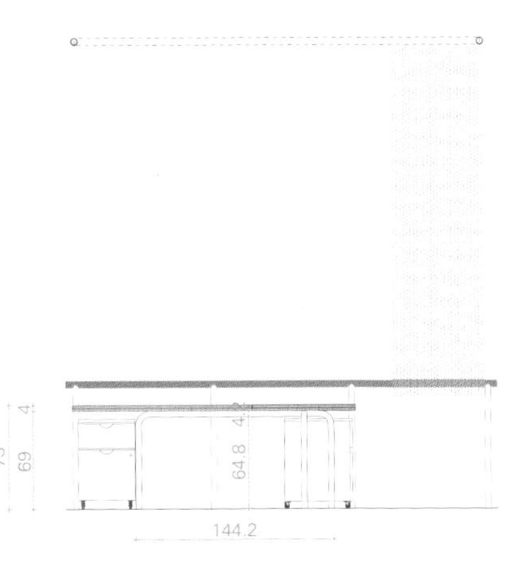

详图2——正视图 detail 2 _ front view

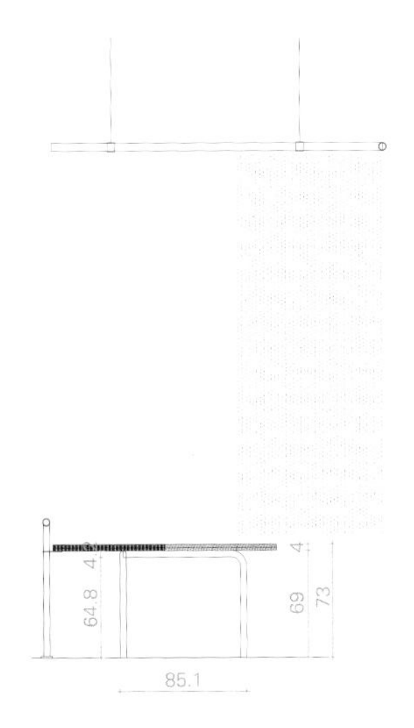

详图2——侧视图 detail 2 _ side view

联合利华北美总部

Unilever North American Headquarters

Perkins and Will

Perkins and Will 建筑事务所将对联合利华整合北美总部的需求做出大胆回应

Perkins and Will deliver a bold response to Unilever's need for a consolidated North American head office

全球消费品公司联合利华决定将其代表1000多个品牌的北美办事处合并为现代总部，此时，对于公司和建筑师来说，风险是很高的。该公司不仅需要将5栋建于20世纪60年代的办公楼改造成一个最先进的工作场所，而且还需要实现碳排放减少50%的目标，这一目标着实让人震惊。

Perkins and Will建筑事务所重新定位联合利华的企业园区，将建筑整合为高度灵活的协作环境。其核心是The Marketplace，这是一个充满活力、光线充足的社区中心，充满了联合利华北美品牌的颜色、风味和个性。

异想天开的室内品牌为千禧一代员工创造了难以忘怀的体验。以健康为本的先进设计策略与卓越的环境工程相结合，确保了超高的性能。因此在纽约市的该场地建起了一座健康、可持续、技术先进的重建大楼。

可持续性

集成智能技术可记录能源数据并实现能源使用自动化，使建筑能够从中学习员工的行为并记住他们的偏好。房间，甚至整个楼层，都可以在不经常使用期间关闭，联合利华员工可以使用自定义应用程序定制其个人照明和视听系统。

福利

一系列空间排布有助于联合利华员工的身心健康。这些设施包括一个提供全方位服务的健身中心、一个提供健康食品的餐厅，专门为处于哺乳期的女性提供的房间、中性浴室和灵活的工作空间。

建筑师还为联合利华的免费员工班车设计了一个方便的乘客接送区，这减少了员工的出行压力。在整个施工过程中，明确规定只使用无毒的建筑产品和材料。

工作场所策略

联合利华品牌还开发了各种各样的其他产品，如Lipton Tea（立顿茶）、Q-Tips（日常护理用品品牌）、Ben & Jerry's（美国仅次于哈根达斯的第二大冰淇淋制造商）和凡士林，现在他们共享公共空间，相互交

流、相互协作。这得益于建筑师与工作场所规划师和战略家的合作，也是联合利华历史上第一次为了合作和创新而整合这些团队。

体验式设计

设计团队将原本散布在园区内五栋建筑的室外庭院围合起来，创造了一个巨大的中庭空间，联合利华员工可以在其中见面、工作、吃饭、娱乐，或者只是放松一下，同时沉浸在他们所代表的公司的品牌和故事中。

灵活的工作场所

员工可以选择工作地点和方式，从而提高工作效率。这已经成为联合利华提高效率、打破筒仓式工作间、吸引并留住纽约市附近千禧一代人才的一种途径。

When global consumer goods company Unilever decided to consolidate its North American offices, representing over a thousand brands, into a modern-day headquarters, the stakes were high for the organization and the architects. Not only did the company need to transform five 1960s office buildings into a single, state-of-the-art workplace, but also, it needed to achieve a staggering 50% reduction in carbon emissions. Perkins and Will repositioned Unilever's corporate campus by consolidating the buildings into a highly flexible, collaborative environment. At its core is The Marketplace, a vibrant, light-filled community hub bursting with the colors, flavors, and personalities of Unilever's North American brands. Whimsical interior branding creates a memorable experience for a millennial workforce. And advanced wellness-based design strategies combined with superior environmental engineering ensure ultra-high performance from the building. The resulting site is a healthy, sustainable, and technologically advanced redevelopment in the New York metropolitan area.

Sustainability

Integrated smart technologies record energy data and auto-

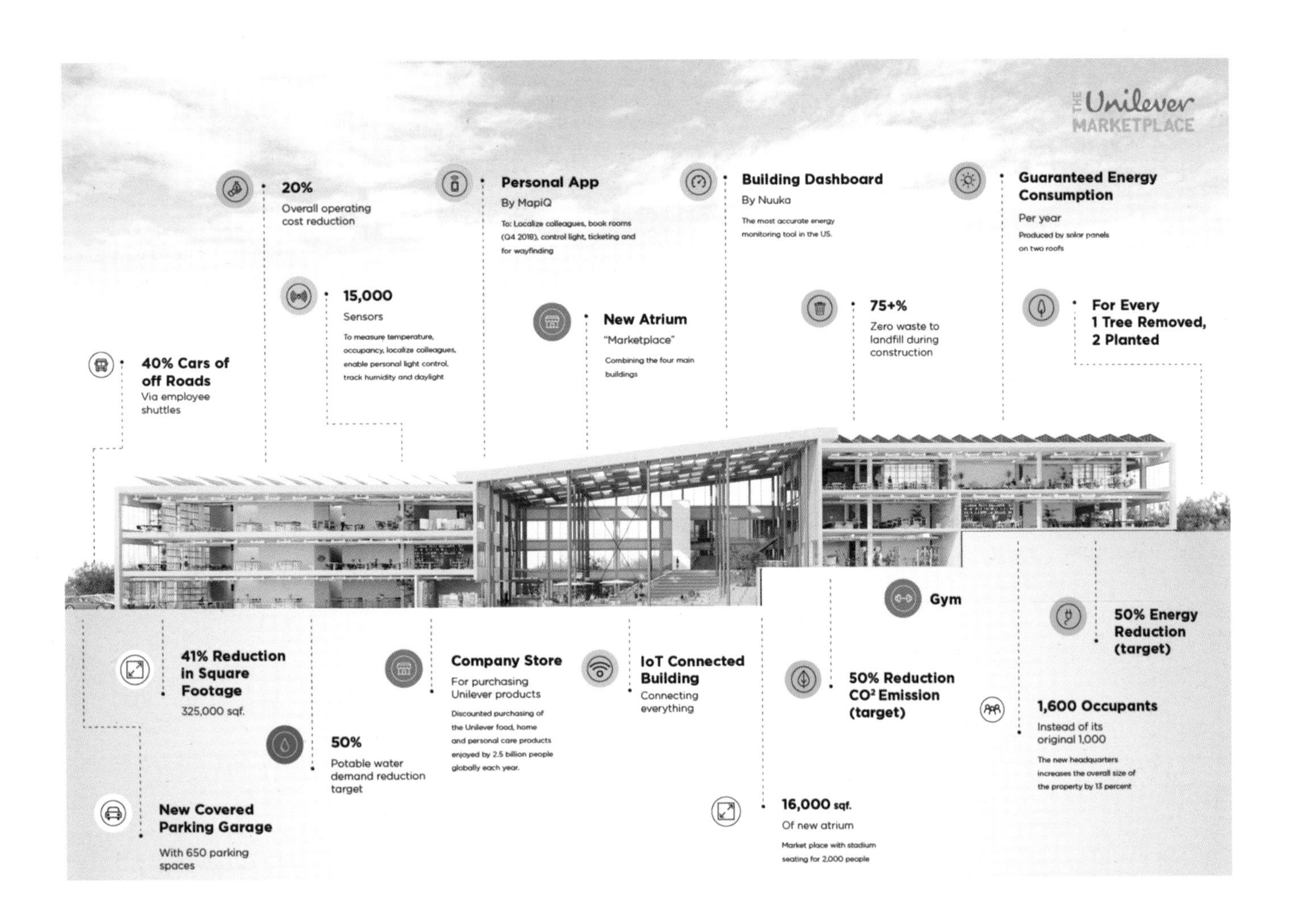

WALMART
SALES OFFICE
ROGERS, AK | 1,268 MILES
UNILEVER HQ
ROTTERDAM NETHERLANDS
3,625 MILES
MARKETPLACE
MARKET
CAFÉ

mate energy use, enabling the building to learn from employee behaviors and remember their preferences. Rooms, and in some cases entire floors, can turn themselves off during periods of infrequent use, and Unilever staff can customize their personal lighting and audiovisual systems with custom apps.

Wellbeing

An array of spaces supports Unilever employees' physical, mental, and emotional health. These include a full-service fitness center, a café serving healthy foods, dedicated rooms for breastfeeding mothers, gender-neutral bathrooms, and agile work spaces.

The architects also integrated into the designs a convenient passenger pick-up and drop-off area for Unilever's free employee shuttle, which reduces travel stress. Throughout the construction only only non-toxic building products and materials were specified.

Workplace Strategy

Employees that work across Unilever's brand verticals, developing products as diverse as Lipton Tea, Q-Tips, Ben & Jerry's and Vaseline, now share common space, interact, and collaborate with one another. This was facilitated by the architects' collaboration with workplace planners and strategists, and is the first time in Unilever's history to consolidate these teams for the purpose of collaboration and innovation.

Experiential Design

By enclosing the outdoor courtyard that had originally separated the five buildings across the campus, the design team created a vast atrium-like space in which Unilever employees can meet, work, dine, play, or simply unwind – all while being immersed in the brands and stories of the organization they represent.

Agile Workplace

Employees choose where and how they wish to work, thereby boosting their productivity. This has become an approach that allows Unilever to increase efficiency, break down silos, and attract and retain millennial talent from nearby New York City.

Unilever

Unilever

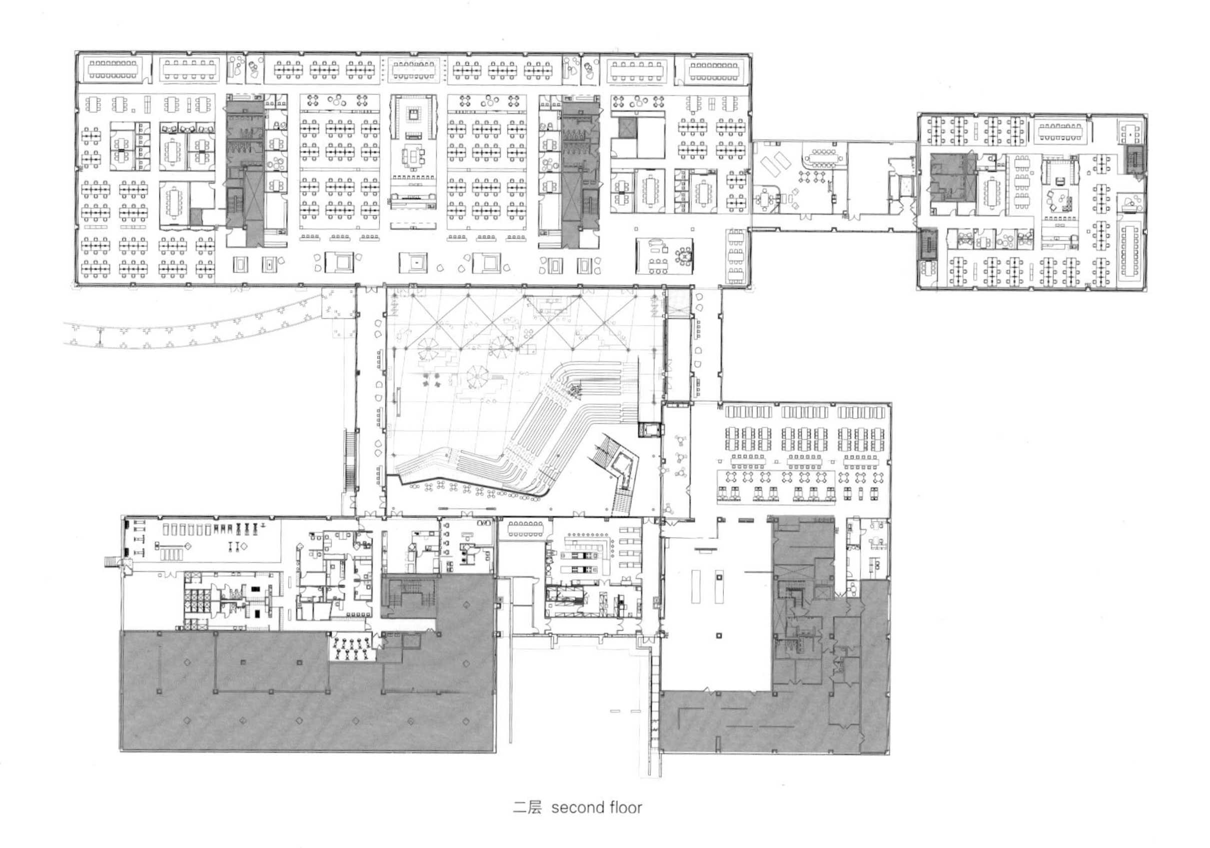

二层 second floor

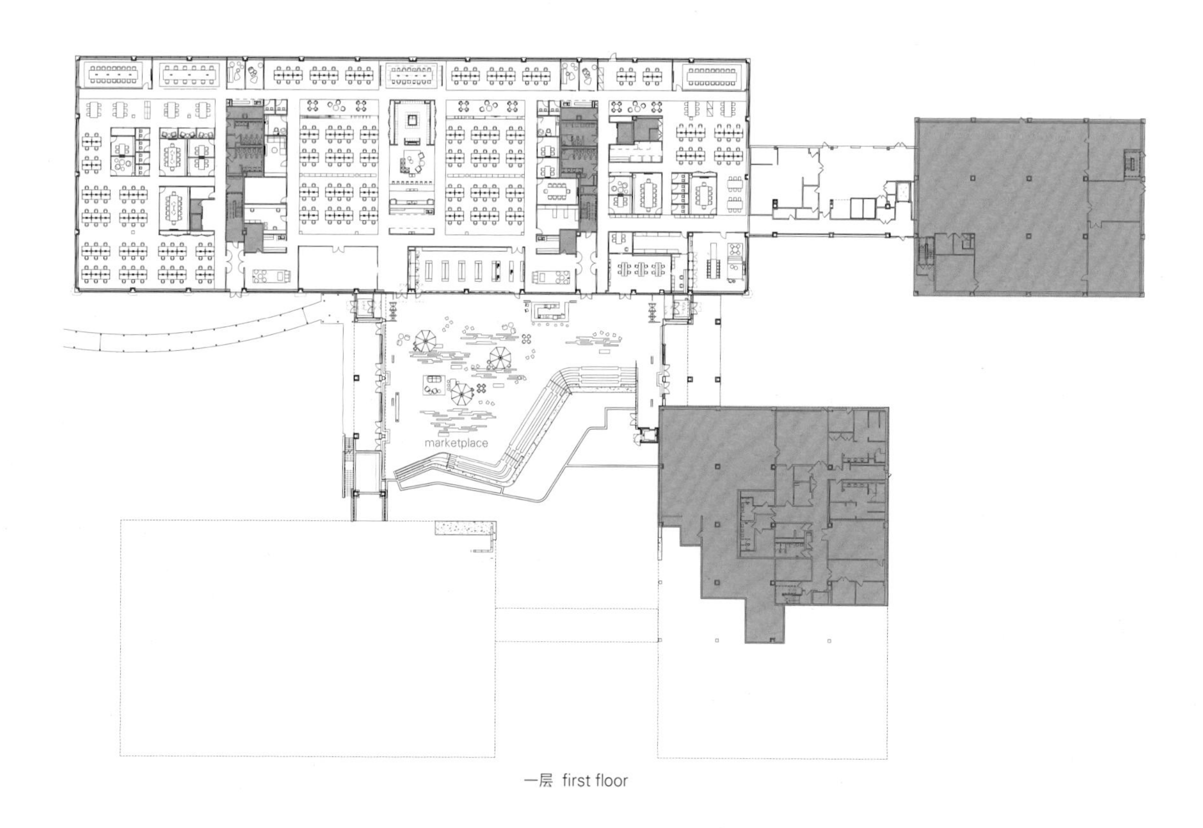

一层 first floor

项目名称：Unilever North American Headquarters / 地点：Englewood Cliffs, New Jersey, USA / 建筑师：Perkins and Will / 客户：Unilever

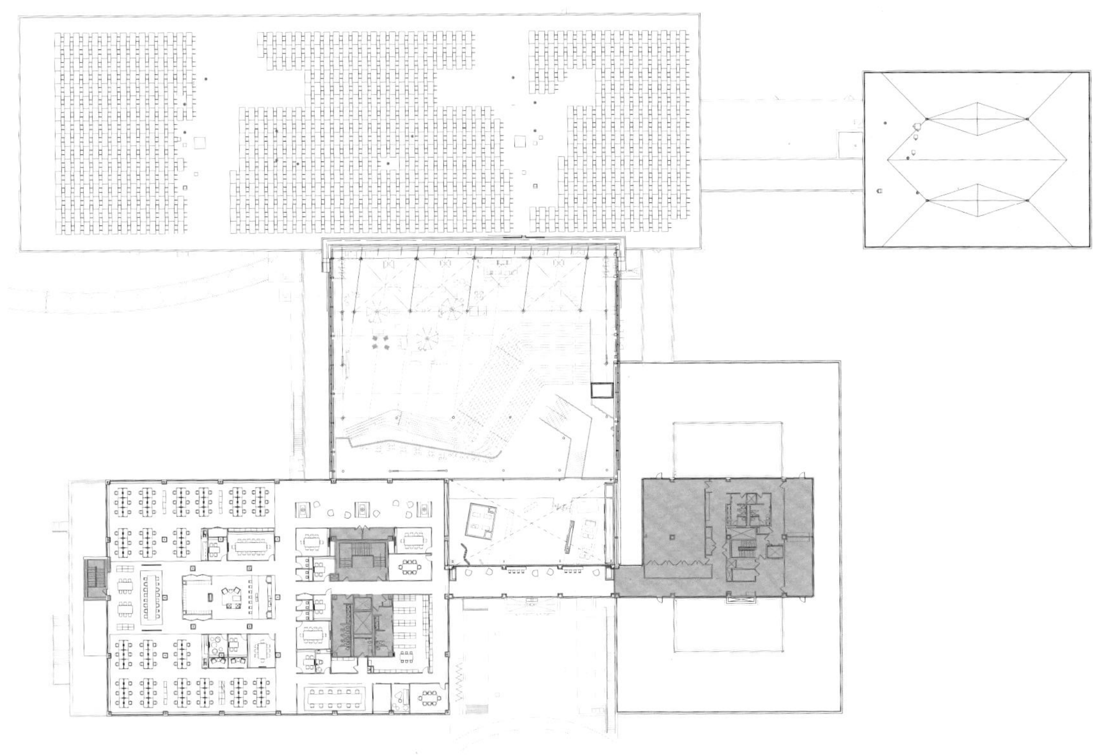

四层 fourth floor

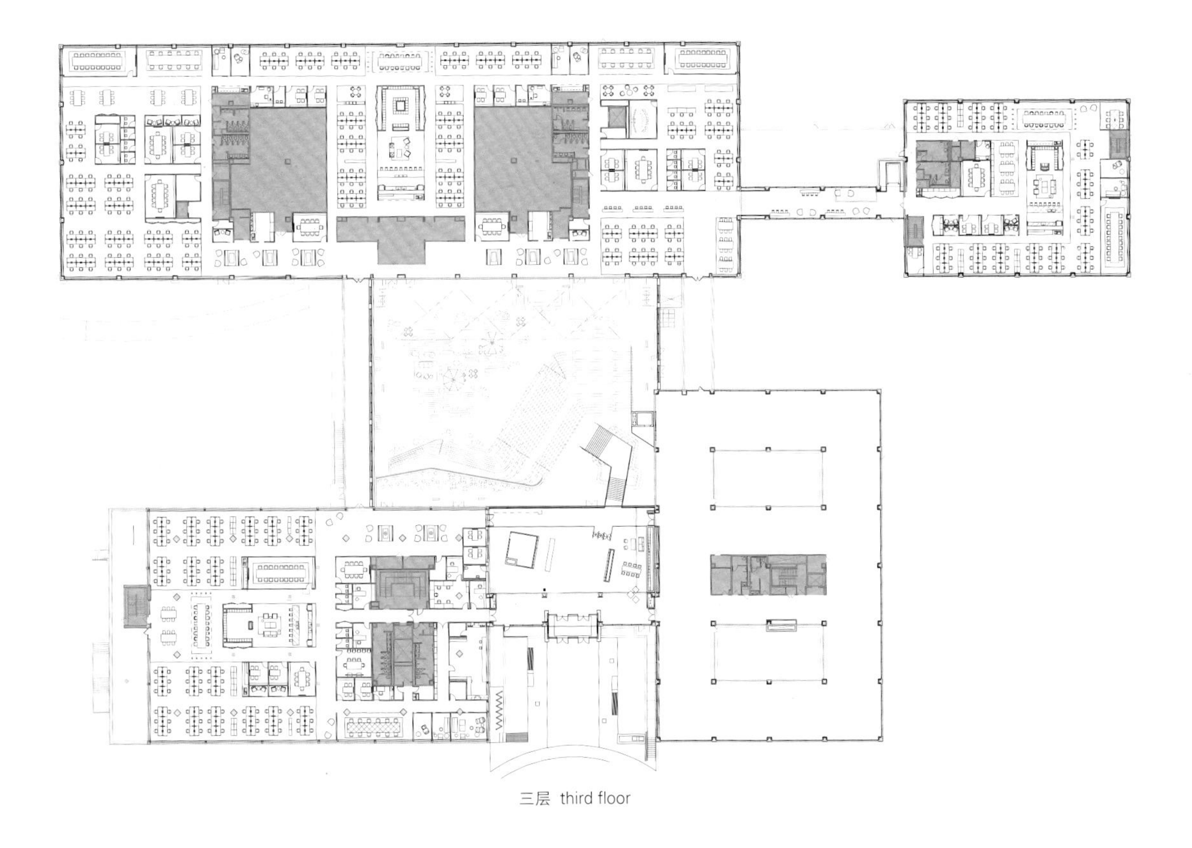

三层 third floor

总建筑面积：27,871m² / 可持续性：LEED Platinum®, WELL Certified Gold / 竣工时间：2018 / 摄影师：©Garrett Rowland (courtesy of the architect)

A billion better lives BEGINS WITH
KNORR, MAILLE, JOYCE FARMS, LANCASTER
Fresh · LOCAL · Seasonal · HEALTHY · Fresh
CAFE
HEALTHY
ENJOY
DELICIOUS
TASTY
LUNCH
FRESH
ORGANIC
HEALTHY
ENJOY
ANOTHER ONE BITES THE CRUST
GOOD FOOD THAT DOES GOOD HELPS YOU THRIVE
MEET THE TEAM

A NEW WAY OF BUSINESS THAT BELIEVES EVERYONE CAN LIVE WELL WITHIN THE LIMITS OF THE PLANET
#ShareAMeal
MAKE SUSTAINABLE LIVING COMMONPLACE
SEND THE BOYS A BOWL OF HOMEMADE-TASTING SOUP
BY MAIL!
LIPTON'S CONTINENTAL NOODLE SOUP
DREAMY-GOOD
Dreamsicle
FOUNDER'S YEARBOOK
FOR SAVING LIFE
LIFEBUOY SOAP
CREAM

大学是公共空间吗?

Univers
as Comm

从1968年到现在，已经过去了半个多世纪，但对于大学应该是什么，大学应该如何在民主、自由的社会建设中定位，仍然是一项艰巨的任务。大学因其培育知识经济的潜力而备受赞誉，同时也因其与商品化的逻辑藕断丝连而受到谴责，而商品化的逻辑甚至把教育也变成了一种货币化的服务，20世纪90年代中期，大学已经被称为是废墟中的机构。然而，这并不一定意味着灾难，因为高等教育阶段在个人塑造过程中仍然是一个关键时期，而此时的个人，即学生以及整个学术界，在年龄、阶级和性别方面日益多样化。因此，大学应该被誉为集体生活的重要场所，高等教育空间新项目的产生为测试这种集体性的变化提供了机

Over half a century has passed since 1968, yet reaching consensus as to what a university should be, and how it should position itself in the construction of democratic, liberal societies, is still a difficult task. Trapped between praise for its potential to nurture a knowledge-based economy, and accusations of complicity with the logics of commodification that have turned even education into a monetized service, the university was already declared an institution in ruins in the mid 1990s. Yet this does not necessarily imply catastrophe, as higher education remains a crucial moment in the formation of individuals, at a time when the latter – the students, but also the academic community at large – are increasingly diversified in terms of age, class and gender. Universities should thus be hailed as important

ties
n Space?

会。这不是一项容易办到和无足轻重的任务，特别是在当时的历史时期，把高等教育与普遍存在的市场心态分离开来变得越来越困难。大学的设计存在这样一种风险，就是一套东西陈陈相因，轻易地演化成一种毫无异议的模式——社会性学习（学习他人的思维和行为）成为中心主旨，在最近的建筑作品中，这种学习特征"蔚然成风"。在此必须发出警告，以避免这些趋势似是而非地被看成是思想的中立和正常化状态，因为在50年前，这一学习模式的初衷是要打破专制的家长式教育的现状。

places to live collectively, with the production of new projects for higher education spaces providing opportunities to test variations on such collectiveness. This is not an easy and a-critical task, especially at a moment in history when dissociating higher education from a pervasive market mentality is becoming more and more difficult. There is a risk that clichés might too easily be reproduced as unquestioned trends in the design of universities – with social learning being the leitmotif celebrated in the vast "learning landscapes" of recent architectural production. A warning must be sent out, to avoid these trends being treated, paradoxically, as the neutralized and normalized versions of ideas which were originally intended to disrupt the status quo of authoritarian, paternalistic education fifty years ago.

大学是公共空间吗?
Universities as Common Space?

Francesco Zuddas

在一个自封知识社会的时代，我们对于继续建设大学并不感到奇怪。知识经济的倡导者提醒我们，事实上，高等教育机构在一个为社会提供服务的行业中扮演着核心角色，这个社会被认为比过去更"聪明"，需要拥有更高脑力技能的劳动力。[1]任何现象都会有人出来进行贬低和批评，对这个主张也是如此。对他们来说，我们的时代应该被更恰当地解释为：基于知识的、有创造性的、创新的等等一切标签都只是在抵消和颠覆它最初的承诺。与过去相比，知识型社会不再是可以不受限制地塑造自我的安全之地，而是由不稳定性、不确定性和扩大的差距所形成的产物。换句话说，这是一个认知资本主义的时代，在这个时代，大众社会已经被"众多的个体"所取代，这样说是因为这些个体永远无法合成为一个整体的"人"，这群个体一直在不断地寻找工作，矛盾的是，由于永久失业，所以他们总是在工作。[2]在这种情况下，大学就是"知识工厂"[3]，产品就是这群个体，他们必须灵活地适应不稳定的生活，因为任何固定的参照点（无论是福利国家，还是跟从前完全相同的工厂）都已不复存在。

我们所描述的当前形势就是在更长的历史轨迹中去捕捉一个短暂的时刻。本章介绍的大学空间项目建立在高校建筑设计和规划的一些既定传统的基础之上：BIG事务所设计的格拉西尔-托尔斯港学院（190页）直接面向周围开阔的景观；Verstas Architects建筑师事务所设计的阿尔托大学（172页），采用了化整为零的模块化设计和内部迷宫式构造；C.F. Moller建筑师事务所设计的医学科研大楼（208页）对垂直度展开探索，完成了一种不太常见的高校建筑设计，实际上本章可以描述为对建筑的历史横向比较。

更广泛地说，这些建筑主题正是当代知识环境的根源，其体系的确立是在20世纪60年代前后。相应地，这个时期会继续引发人们对当前建筑话语的兴趣，作为一个实验性时代，我们仍然可以从中学习到东西。20世纪60年代，高校设计推动着建筑学经历了一段近代史上最微妙的反思时期。大家最喜欢通过大学这个领域考察从"以前"到"后来"的过渡，产生了各种各样的叙述——从"现代主义"说到"后现代主义"，从"国际现代建筑协会"说到"十次小组"，从实用建筑主义说到公众参与，等等。

与此同时，从50年前开始，新的比喻词语重新定义了建筑的词汇，也就是谈论空间的方式。除了回归街头的场景（无论是在地面上还是在"空中"），灵活的、适应性强的、非正式的和令人意想不到的建筑词汇，作为革新的媒介出现，带领我们学习并超越现代建筑运

In a self-declared age of knowledge-based societies, it is no surprise that we keep on building universities. The advocates of the knowledge economy remind us that, indeed, higher education institutions play a central role in an industry servicing a society supposedly more "intelligent" than in past times, which requires a workforce with higher cerebral skills[1]. As with any phenomenon, the present one also has its detractors and critics. For them, our time should be more aptly interpreted as one in which any labelling – knowledge-based, creative, innovative, etc. – is merely the neutralized and inverted version of its original promise. Instead of being havens of freer self-formation compared to past times, knowledge-based societies are shaped by precariousness, uncertainty, and enhanced disparity. Alternatively termed, this is the age of cognitive capitalism, where mass society has given way to a "multitude of individualities" that never reach synthesis as a "people" – a multitude made of subjects in the continuous search for employment, paradoxically always working because permanently unemployed[2]. In such scenarios, universities are "factories of knowledge"[3], production plants of individuals that have to be flexible to adapt to a life of instability, because any fixed point of reference – from the Welfare State to the self-same factory of the olden days – has gone.

The present situation captures a fleeting moment within a longer historical trajectory. The projects for university spaces presented in these pages build upon some established traditions of design and planning for higher education: the building set in direct confrontation with an open landscape, as represented by BIG's design for Glasir Tórshavn College (p.190); an exercise in modularity and interior maze-creation, in the project for Aalto University by Vertas Architects (p.172); and an exploration of verticality in C.F. Moller Architects' Maersk Tower (p.208), which deals with a less common line of higher education design (which could in fact rather be described as a history of horizontality).

These architectural themes and, more generally speaking, the very roots of the contemporary condition of knowledge and its institutions can be located in and around the 1960s. It is a period that, in turn, continues to attract the interest of current architectural discourse as an epoch of experimentation from which we can still learn. During the 1960s, university design drove architecture through one of the most delicate moments of rethinking in its recent history. Universities became a favorite ground to test the passage from a "before" to an "after" that has been variously narrated – modernism to post-modernism, CIAM to Team X, functionalism to participation, etc.

Concomitantly, starting five decades ago, new tropes renovated the very vocabulary of architecture – that is, the way in which space is talked about. Besides the return on the scene of the street (be it on the ground or "in the air") the flexible, the adaptable, the informal and the unexpected appeared as the agents of renovation to take us beyond, yet learn from, the teachings of the Modern Movement. Taking advantage of a demand for higher education spaces at an unprecedented scale – both in numbers and in the size of the new campuses and academic buildings built throughout

动的教义。在整个20世纪的六七十年代，对高等教育的空间需求量——无论是在新建校园和科研楼的数量上还是尺寸上，都达到了前所未有的规模，利用这种需求，这些术语经由当时的主要建筑师之口表述出来，反过来又为他们提供了千载难逢的机会，让他们在建筑史上留下自己的足迹。

专门有一章来谈论大学的设计和规划，其中列举了柏林自由大学、英国一些新建的大学、加拿大和意大利的巨型结构方案，散布在西方殖民地的诸多新校区，以及其他许多关于学术空间的设想，这些设想作为可能塑造出具有更平等学习机会的新的民主社会动因而得到了讨论和捍卫。[4]

当代大学空间的建造在陈述上同样是气势恢宏的，如果我们回顾过去五十年来高校建筑设计的演变，就会看到我们是如何达到这样一种局面的，这种局面总是以最大的信念和坚持为特征，坚持相信大学内部公共空间创造学习机会的潜力，远远超出了学者坐在书桌旁或学生排成一排面对讲师的传统静态形象。

本章以这种方式解读各个项目，这些项目都基于同样的理念，那就是将大学当作公共空间。由于字数原因，请允许我们聚焦其中一个项目，以便能够更全面地解释当前学术空间的产生。

格拉西尔-托尔斯港学院坐落在能俯瞰法罗群岛海面的山丘上。一系列堆叠的线性盒状建筑从一个公共中心当中分离出来。或者，从另一个角度看，则是一个中央公共区域在建筑的每一层展开，从不同的方向与周围环境接触。这种对活力的追求在项目的照片中得到了加强。这些照片遵循了一种叙事的路线，这种叙事路线在近期对任何研究高等教育空间演变的人来说都是熟悉的。这些图像以动态的形式捕捉了建筑的"躯体"，通过从城镇到建筑的通道，进一步描绘了建筑内部的运动。

这段永无止境的旅程描绘出了一种"孤独感"，当一个人在旅途中独自行走时可能会产生一种孤独感，但这里的孤独感却用来描述大学所呈现出来的一种元素，这还是十分少见的。联想，及其对于社会性学习的假设，是一个空间的主体，个体在这个空间中以临时的形态聚集在一起，而这些临时形态总是存在于更广阔的社区视角或观众视角中。在这个空间里，学习（这个词在当代大学里的有效性已经开始受到质疑[5]）和打情骂俏成为相似的空间实践，三对男女学生在阶梯教室里相互微笑的照片成为当代大学的典范，就像在过去的大学讲堂里人山人海的照片一样。

the 1960s and 1970s – those terms structured the narratives of some of the major architects of the time, in turn providing them with the opportunities of a lifetime to leave their trace on the books of architectural history.

A specific chapter on university design and planning was written up, its pages illustrated by the likes of the Berlin Free University, the British new universities, the Canadian and Italian mega-structural proposals, the many new campuses scattered around the western colonial territories, and many other projected visions of academic spaces that were discussed and defended as the possible agents to shape new democratic societies along with more equal learning opportunities[4].

The contemporary production of university spaces is no less heroic in its statements, and if we look at the evolution of higher education design over the last fifty years we see how we have come to a situation which is invariably characterized by utmost faith in, and insistence on, the potential of the common spaces inside universities to forge learning opportunities well beyond the traditional, static image of the scholar sitting at a desk or of rows of students facing a lecturer.

Read in this way, the projects presented here are variations on a single idea – the university as common space. Focusing, for reasons of word space, on one of those projects, allows us to interpret more generally the current production of academic space.

The Glasir Tórshavn College sits on a hill overlooking the sea in the Faroe Islands. A series of stacked linear boxes depart tangentially from a common center – or, seen the other way around, a central communal area gets unfolded at each level of the building to reach out to the surroundings in different directions. Such strive for dynamism is reinforced by the photos chosen to depict the project, which follow a narrative line that has become familiar for anyone studying the evolution of higher education space in recent times. In these images, bodies are captured in movement, channeled through the pathways that lead from the town to the building, and further depicted on the move inside it.

This never-ending journey portrays isolation as something that is possible when on the move (a person walking alone) but rarely presented as a component of what a university is. Association, and its postulate of social learning, are the protagonists of a space where individuals assemble in temporary configurations that are always set in the plain view of a wider community or audience. Here, studying (that old word whose validity in the contemporary university has started to be questioned[5]) and flirting are similar spatial practices, the image of three boys and three girls' complicit mutual smiles across the stepped interiors being as paradigmatic of the contemporary university as a photo of a crowded lecture auditorium would have been for a university of the past.

The auditorium or, more generally, the lecture hall and the classroom, have disappeared from the way that universities are talked about now, in words and images that relentlessly orient their gaze on the spaces "in-between". Such disappearance

大礼堂，或者更广泛地说，讲堂和教室，已经从现在谈论大学的方式中消失了，在文字和图片中出现的都是"中间"空间。如果所讨论的空间（可能有人会怀疑，但它也是唯一从建筑学的角度进行过深思熟虑的空间）是非特定的公共空间，而不是"正式的"教学空间，那么这种消失就引发了"介乎什么中间"的问题，这自然是个合理的问题。当然，任何熟悉学术界寻常做法的人都知道，讲堂和教室并没有消失，依然是由墙壁区分的空间，即使是有五花八门的学习方法，但板书和讲授的教育仍然是知识交流的核心手段。问题仍然在于，为什么这些空间从来没有被展示或描述过，就好像它们被一些原罪所腐蚀了，这让建筑师羞于承认该空间在学习过程中的作用。相反，试图写一段教室的历史可能会揭示出一个丰富的思想宝库，就像今天高等教育叙事中过度描绘的公共领域和"学习场景"一样。

高等教育的现状似乎与50年前人们所期望的那样相反，当时世界各地的大学都受到了大规模抗议浪潮的冲击，当时的建筑师正在设想新的学术环境。尽管20世纪60年代呼吁非家长式的教育，个性化的学习途径，跨越种族、性别和社会阶层的更平等的教育机会，以及高等教育研究与社会需求更紧密的联系，但是当代商业化和企业化的大学通常仍然是一台控制机器，它已经取代了早期的纪律性思维模式，取而代之的是基于监督性的思维模式。

用一种稍微不同的方式来解释就是，现在的大学也许更适合20世纪60年代愿景中描述的中性化、规范化的样子，这在当时被描述为对现状的破坏。因此，当时流行的任何概念（即终身学习、非正式性、灵活性），作为从精英主义和对高等教育的过时理解中获得解放的媒介，都已经成为大学保持自上而下控制的工具。糟糕的是，当代大学里存在着把任何事情都视为一种经济交易的做法，把知识（广义的教学和研究）当成了市场上购买的商品[6]。

在这样严重的情况下，当前建成的大学建筑很容易成为受人怀疑的对象。基于上面提到的终身学习、非正式性和灵活性，目前的项目在多大程度上只不过是一个更庞大的商品化知识和认知资本主义系统的共同空间机制？[7]

在目前仍有新的大学空间出现的情况下，这些问题尚在探索，让我们把它们留给读者，现在可以将注意力转向这一章要介绍的项

stimulates the legitimate question of "in-between what?", if the only spaces discussed (but, one might suspect, also the only ones really thought-through architecturally) are the non-specific, common spaces other than the "formal" ones of instruction. Of course, anyone acquainted with the everyday practices of academia knows that the lecture hall and the classroom have not disappeared, and that they are still defined by walls because, despite the informal learning rhetoric, chalk-and-talk remains a core means of knowledge exchange. The question remains as to why those spaces are never shown or described – as if they are corrupted by some original sin that makes architects ashamed at conceding them a role in the learning process. Conversely, trying to write a history of the classroom might unveil as rich a repository of ideas as the obsessively portrayed common areas and "learning landscapes" of today's higher education narratives.

The present condition of higher education appears as the antonym of what was hoped for fifty years ago, when a vast wave of protests shook universities around the world at the very time when architects were envisaging new academic environments. Despite the 1960s calls for non-paternalist education, paths to learning that can be personalized, more equal educational opportunities across races, genders and social classes, and closer ties of higher education studies to the needs of society, the contemporary commodified and corporate university is often still a controlling machine that has substituted its earlier disciplinary mentality to one based on surveillance.

Interpreted in a slightly different way, the current university might perhaps more aptly be described as the neutralized, normalized version of the 1960s vision that was, at the time, depicted as a disruption of the status quo. As such, any of the concepts that gained currency then, as the mediators to achieve liberation from an elitist and old-fashioned understanding of higher education, (namely life-long learning, informality, flexibility), have become none other than the very instruments by which universities retain top-down control. The situation is made even worse by the fact that anything in the contemporary university is treated as an economic transaction, using knowledge (in the broad sense of teaching and research) as a commodity to be purchased on the market.[6]

Set within such a critical scenario, the current production of university architecture can easily be the target of suspicion. To what extent are the current projects – relentlessly based on and promoted according to those very categories recalled above, of life-long learning, informality and flexibility – nothing more than the complicit spatial mechanisms of a vaster system of commodified knowledge and cognitive capitalism?[7]

Leaving to the reader such questions as a possible line of inquiry into the current, still conspicuous, construction of new spaces for higher education, we can turn our attention to the projects presented in these pages and try to observe them in their own right as media of spatial reasoning, and as the latest products of a most beloved brief for experimentation by architects. In other words, we can momentarily suspend the invitation to consider these projects as cliché-ridden designs, something that their architects' explanatory texts would suggest (in fact, such criticism applies as easily to almost any project description, making them perfect fodder for those with a conspiratorial mentality). This means also

目，作为表现空间妥当性的手段，同时也是建筑师最喜欢的新式实验的结果，我们将站在他们的角度去观察这些作品。换句话说，我们可以暂时不考虑将这些项目视为陈旧的设计，虽然它们的建筑师给出的说明文本可能会让人看出这种端倪。（事实上，这样的批评很容易适用于几乎所有的项目描述，使它们成为阴谋论者的完美素材）。这也意味着要拒绝使用以"这些项目的血管中流淌的是商业"为论调的诠释（其实非常明显，例如，奥塔涅米的阿尔托大学新楼的设计师陈述：这所商学院作为艺术的伙伴呈现在一种混合结构中，据称这种混合结构是一种创新的大学理念）。

最后，我们还可以试着暂时搁置，或者最好放弃任何决定性的态度，这种态度过于根深蒂固，通常认为高等教育机构是"行为机器"，能够影响或指导他们内部的生活。

因此，我们可以暂时忘记"社会化中心"，内化的"学习之路""意外相遇的平台"以及其他反复出现的解释，这些解释充斥了许多建筑师的项目描述。取而代之的是，让我们更简单地考虑其实质，以表明我们仍然需要大学作为集体生活的重要机会。或者，用比尔·雷丁斯的话说，尽管我们可能同意或不同意大学的"理念"，但我们还是将大学视为我们共同生活的许多地方中的一个[8]。

暂缓对建筑设计与我们基本制度商品化之间关系的探究，并不意味着放弃这种探究。然而，我们可以因此先不去纠结这种消极性可能带来的限制，并承认任何空间——即使是最受控制的空间——都可能被重新划分为更好的领域，并且采用一种更好的与世界建立联系的方式，而不是单一维度的方式。从这个角度来看，大学空间的持续建造可以很受欢迎，而且应该受到欢迎，因为这是对未来社会生产的一个关键贡献。未来社会能够将协同合作作为核心价值观加以保留，无论当前的高等教育形势有多么像是在利用每一个机会来打击我们的这种乐观情绪。

正如今天，我们想知道如何处理福利国家及其空间产品的残余——包括20世纪下半叶建造的许多大学——我们迟早会面临一个类似的问题，是否及何时能够挺过目前的晚期资本主义阶段。同时，我们应该把所有新建成的高等教育空间当作礼物，永远不要批判性地接受，而是从个人和集体的角度进行分配和协商。

resisting interpretation based on how "business" is the blood running through the veins of these projects (most explicitly when a business school is being presented as the companion of the arts in a mix that claims to be an innovative idea of university, as stated by the designers of the new building for the Aalto University in Otaniemi).

Finally, we might also try putting on hold – or, perhaps better, renouncing – any deterministic attitude that is excessively embedded in a common way of approaching higher education institutions as "behavioral machines" capable of influencing – or directing? – life inside their walls.

So, we can temporarily forget about the "hubs of socialization", the interiorized "streets of learning", the "platforms of unexpected encounters", and the other recurring explanations that populate many architects' project descriptions. Instead, let us more simply consider the projects in their nature of instances showing how we still need universities as important opportunities for collective life. Or, to use the words of Bill Readings, let us approach universities as one among many other places where we live together, despite any possible corruption of the "idea" of the university – on which we can agree or disagree[8].

Putting on hold an inquiry into the relations between architectural design and its complicity in the commodification of our basic institutions does not mean abandoning such inquiry. Yet, it allows us to loosen the possible ties of negativity and concede that any space – even the most controlled one – might be re-territorialized to better ends and according to a way of relating to the world that is less mono-dimensional. Approached under this light, the continuous production of university spaces can, and should, be welcomed as a crucial contribution to the production of future societies capable of retaining collaboration and cooperation as core values, no matter how much the current higher education climate seems to take every opportunity to sink such optimism.

Just as, today, we wonder about what to do with the remnants of the Welfare State and its spatial products – including the many universities built in the second half of the last century – we will sooner or later be confronted with a similar question if and when the current stage of late capitalism is overcome. In the meantime, we should hail any new space for higher education as a gift, never to be critically received, but personally and collectively appropriated and negotiated.

1. See Ali Madanipour, *Knowledge Economy and the City: Spaces of Knowledge* (New York: Routledge, 2011).
2. See in particular Paolo Virno, *A Grammar of the Multitude* (Los Angeles: Semiotext, 2004).
3. See Gerald Raunig, *Factories of Knowledge: Industries of Creativity* (Los Angeles: Semiotext, 2013).
4. Stefan Muthesius, *The Postwar University: Utopianist Campus and College* (New Haven, CT, and London: Yale University Press, 2000); Francesco Zuddas, *The University as a Settlement Principle: Territorialising Knowledge in Late 1960s Italy* (Oxon and New York: Routledge, 2020).
5. See Giorgio Agamben, 'Studenti' (2017). Retrieved at: https://www.quodlibet.it/giorgio-agamben-studenti.
6. See Derek Bok, *Universities in the Marketplace: The Commercialization of Higher Education* (Princeton, NJ: Princeton University Press, 2003). For more recent criticism on the commodification of universities see Raewyn Connell, *The Good University: What Universities Actually Do and Why It's Time for Radical Change* (London: Zed Books, 2019).
7. See Douglas Spencer, *The Architecture of Neoliberalism: How Contemporary Architecture Became an Instrument of Control and Compliance* (New York: Bloomsbury, 2016).
8. Bill Readings, *The University in Ruins* (Cambridge, MA: Harvard University Press, 1996).

A!
A!
Brooklyn
Cafe

芬兰阿尔托大学瓦雷大楼
Aalto University Väre Building

Verstas Architects

新的瓦雷大楼与现代主义大师阿尔托最初的大学规划完美契合

New Väre building fits seamlessly into modernist master Aalto's original plans for the university

阿尔托大学瓦雷大楼是阿尔托大学艺术、设计、建筑学、商学院的新家。瓦雷在英语中有波纹和微光的意思。大楼坐落在芬兰埃斯波的奥塔涅米校区，这里拥有蜚声世界的现代主义建筑。瓦雷的落成为校园注入了新的活力，并对阿尔托大学的转型、商科和艺术设计以及技术科学等学院的合并以及奥塔涅米区域的集中化和多样化起到了关键的作用，使奥塔涅米区作为芬兰大都市区的一部分持续蓬勃发展。

瓦雷大楼与阿尔瓦·阿尔托和艾莉莎·阿尔托在20世纪60年代设计的主楼及图书馆相邻，因此它还面临着一个挑战，就是新大楼要与这位备受尊崇的现代大师塑造的建筑环境相适应。艾诺·阿尔托和阿尔瓦·阿尔托在1948年对拥有精致景观的校园进行了整体规划。瓦雷大楼在与校园环境形成对话的同时，也从本质上改变了校园的动态。

设计方案引入了一个全新的中心广场，将新教学楼与旧教学楼连接起来，提供了通往地下和瓦雷大楼内部的购物中心的新入口。占地7300m²的购物中心包含餐厅、咖啡厅、健身房和零售场所，因此，瓦雷大楼与新广场共同作为休闲集会的理想场所，还为1850名学生和350名教职员工带来了一系列跨学科的互动空间。

这座建筑使传统的庭院式校园建筑类型与当代的学习和教育方式相适应。简单的网格布局提供了交互且灵活的单元式工作空间。经过谨慎规划的空间序列围绕着天窗下的中庭层层展开。在中庭可以看到所有的楼层以及连接它们的楼梯。透明的墙壁将项目空间中的创意展示出来，底层的大厅也完全向公众开放。

平面布局和模块设计将大规模的瓦雷项目划分为多个较小的单元，使它能够适应公园式校园的环境和规模。两个主立面的轮廓从既有的主楼轮廓当中衍生。手工铺砌的红砖延续了阿尔托在材料运用上的纯熟技艺；印花玻璃则为建筑赋予了现代气息。在其内部，木质表面突出了空间序列在建筑中的流动。设计学院和商学院的室内空间各具

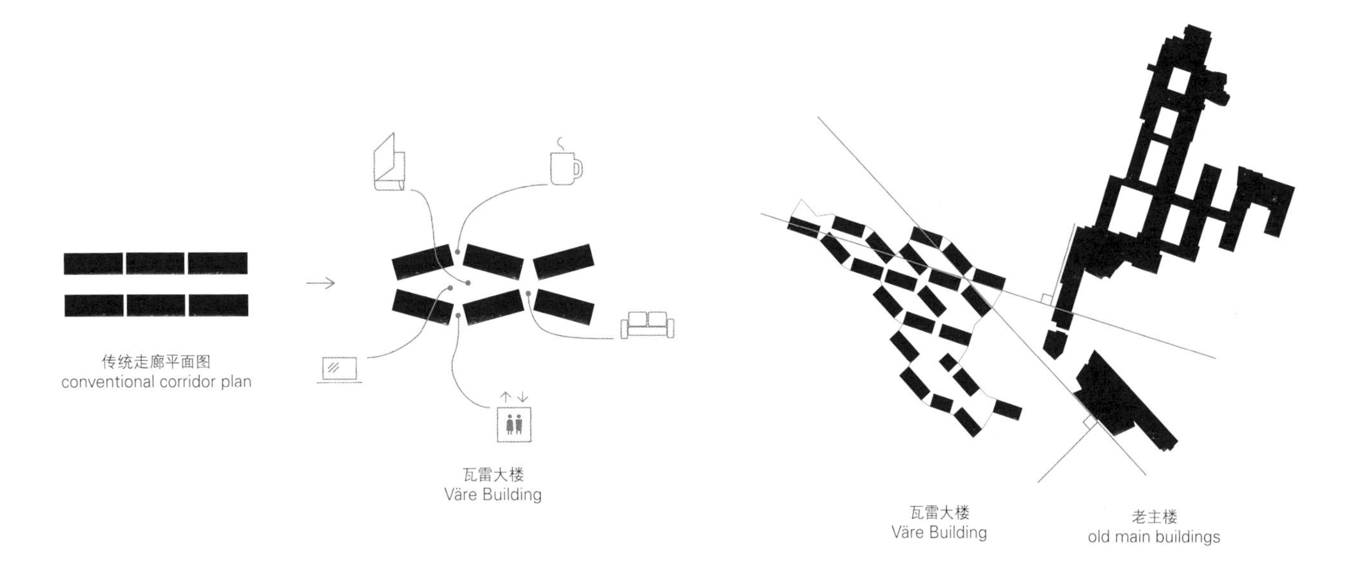
传统走廊平面图
conventional corridor plan
瓦雷大楼
Väre Building
瓦雷大楼
Väre Building
老主楼
old main buildings

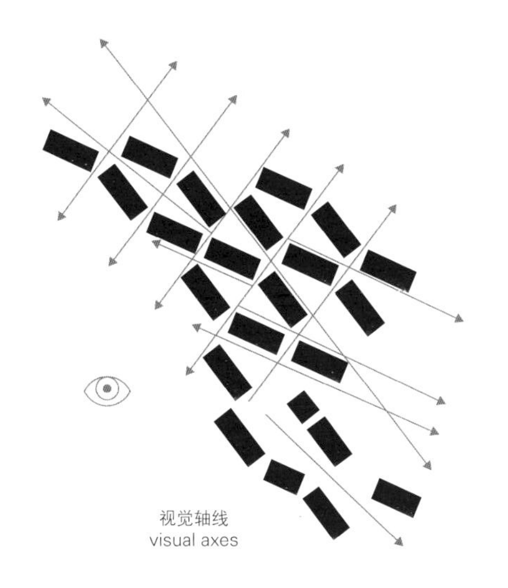
视觉轴线
visual axes

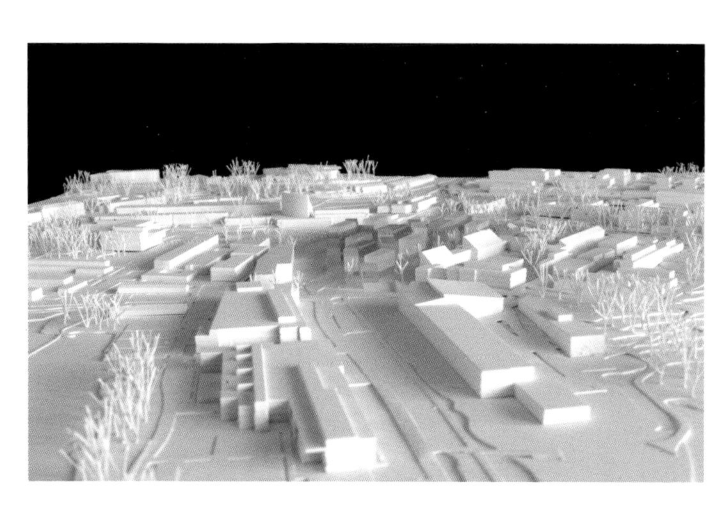

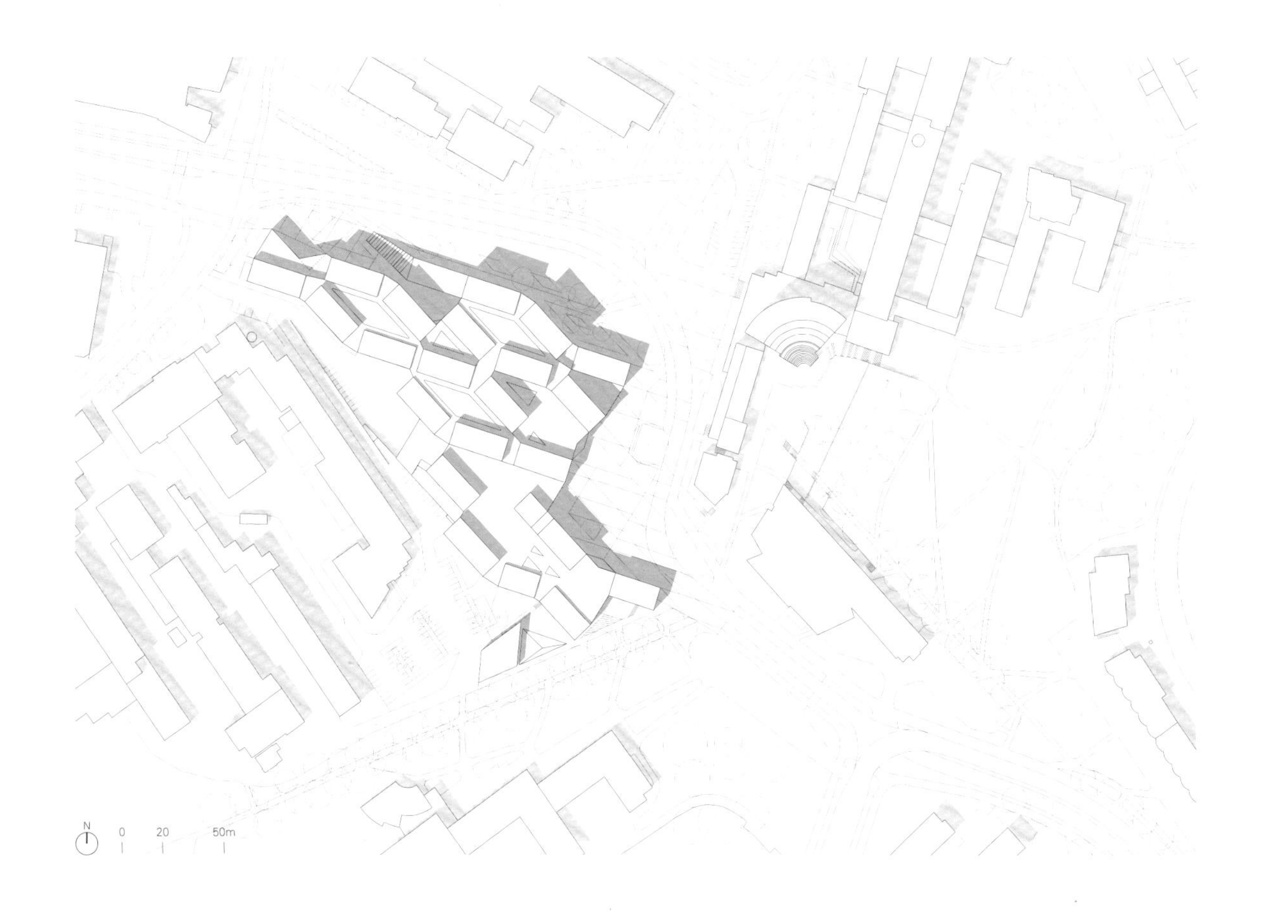
N
0
20
50m

特色。艺术、设计和建筑学院拥有一个自承重的、形如雕塑的钢质楼梯，在中性的背景下显得格外引人瞩目；商学院则采用了石灰石、橡木和温暖的金属色调，营造出温馨而又不失庄严的氛围。

瓦雷大楼具有环保、节能、节水的特点。200个自行车停车场和更衣室鼓励员工骑自行车来上班。90%的热量消耗由地源热泵提供，屋顶的太阳能电池板则可以提供约6%的电能。通风系统可根据需求进行调节，并且拥有热回收功能。照明系统同样可控。此外，还有75%的施工废料得到了回收。

The Aalto University Väre Building is the new home of the Aalto University School of Art, Design and Architecture and the School of Business. Väre, which means ripple or glimmer in English, brings together the schools of design and business in a globally unique way. It is situated in the Otaniemi campus in Espoo, Finland, world-famous of its modernist architecture, and introduces a significant new dynamic to the campus. The Väre building plays a key role both in the transformation of the Aalto University, a merger of earlier universities of business, arts and design and technical sciences, and in the densification and diversification of the Otaniemi district as a part of the growing Finnish metropolitan region.
The Väre building is adjacent to the earlier main building and library designed by Alvar and Elissa Aalto in the 1960s. The project had therefore the added challenge of fitting into the context of highly respected modern master. Aino and Alvar Aalto laid out the whole original masterplan for the carefully landscaped campus in 1948. Väre is in a dialogue this composition, while it also changes the dynamic of the campus essentially.

项目名称：Aalto University Väre Building
地点：Otaniementie 14, Otaniemi 02150, Espoo, Finland
建筑师、景观设计师：Verstas Architects
主持建筑师：Jussi Palva (partner-in-charge), Väinö Nikkilä, Riina Palva, Ilkka Salminen
项目经理：Mikko Rossi
项目团队：Aapo Airas, Aino Airas, Heidi Antikainen, Otto Autio, Anna Björn, Kari Holopainen, Erik Huhtamies, Anna Juhola, Jukka Kangasniemi, Saara Kantele, Pyry Kantonen, Sari Kukkasniemi, Emma Kuokka, Oksana Lebedeva, Johanna Mustonen, Ville Nurkka, Arto Ollila, Milla Parkkali, Miguel Pereira, Pasi Piironen, Teemu Pirinen, Anna Puisto, Aleksi Räihä, Pekka Salminen, Lauri Salo, Katri Salonen, Juhani Suikki, Maiju Suomi, Anniina Taivainen, Tuulikki Tanska, Ilkka Törmä - design team; Soile Heikkinen - landscape architect; Tero Hirvonen, Karola Sahi - interior architects
项目管理顾问：A-insinöörit Rakennuttaminen
施工经理：SRV Construction Ltd / 结构和暖通空调工程师：SitoWise / 基础工程师：A-insinöörit Suunnittelu / 电气工程师：Rejlers /
消防工程师：L2 Paloturvallisuus / 土木工程师：Ramboll / 能源顾问：Green Building Partners / 客户：ACRE Aalto University Campus & Real Estate /
总建筑面积：45,400m² (The Väre Building - 27,400m²; School of Business - 10,600m²; A Bloc - 7,300m²) / 体积：212,300m³ / 造价：Approximately 140 Meur
设计竞赛时间：2012.4—2013.8 / 城市规划批准时间：2015.6
开放时间：The Väre Building - 2018.9; School of Business - 2019.2; A Bloc - 2018.8
摄影师：©Andreas Meichsner (courtesy of the architect) - p.172~173, p.174, p.176, p.178[top-left, top-right], p.179[upper], p.180~181, p.184, p.185; ©Mika Huisman (courtesy of the architect) - p.186, p.187, p.188, p.189; ©Tuomas Uusheimo (courtesy of the architect) - p.178[bottom], p.179[lower]

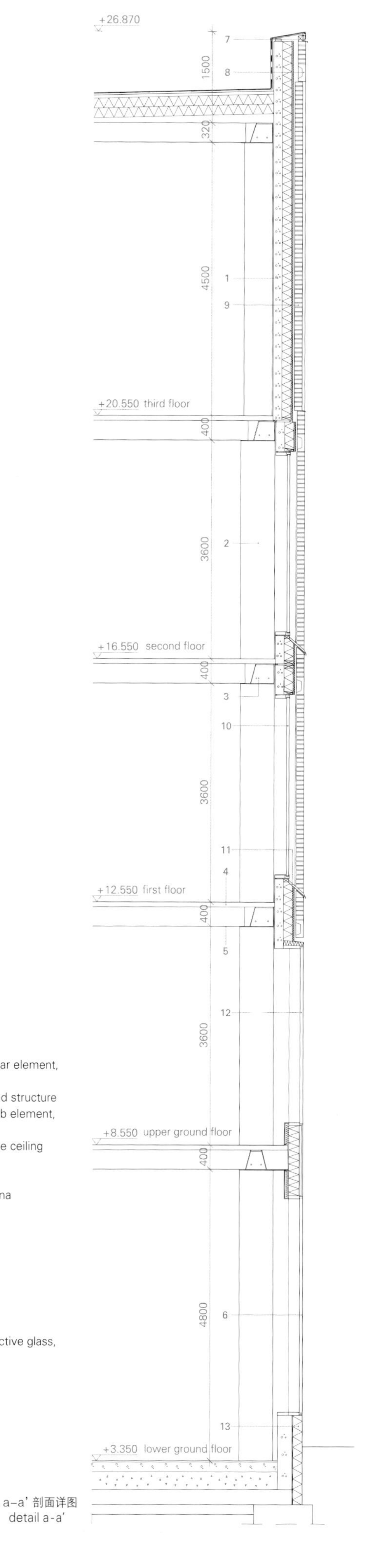

1. prefabricated concrete wall element, painted
2. prefabricated concrete pillar element, concrete cast surface
3. steel beam with connected structure
4. prefabricated concrete slab element, polished concrete surface
5. acoustic board glued to the ceiling
6. structured aluminum glass wall, painted
7. copper eaves, natural patina
8. recessed masonry
9. brickwork laid in situ
10. iron oxide-free solar protective glass, glazed aluminum window on the interior side
11. copper window sheets with natural patina
12. iron oxide-free solar protective glass, silk-screen printed pattern
13. plinth, painted metal

a-a' 剖面详图
detail a-a'

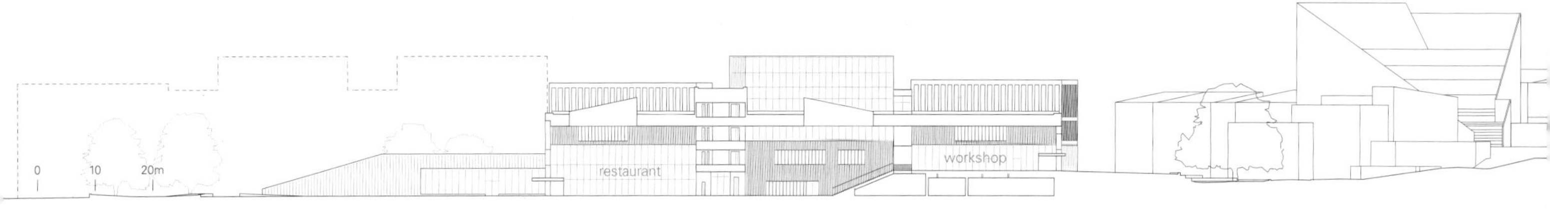

A–A' 剖面图 section A-A'

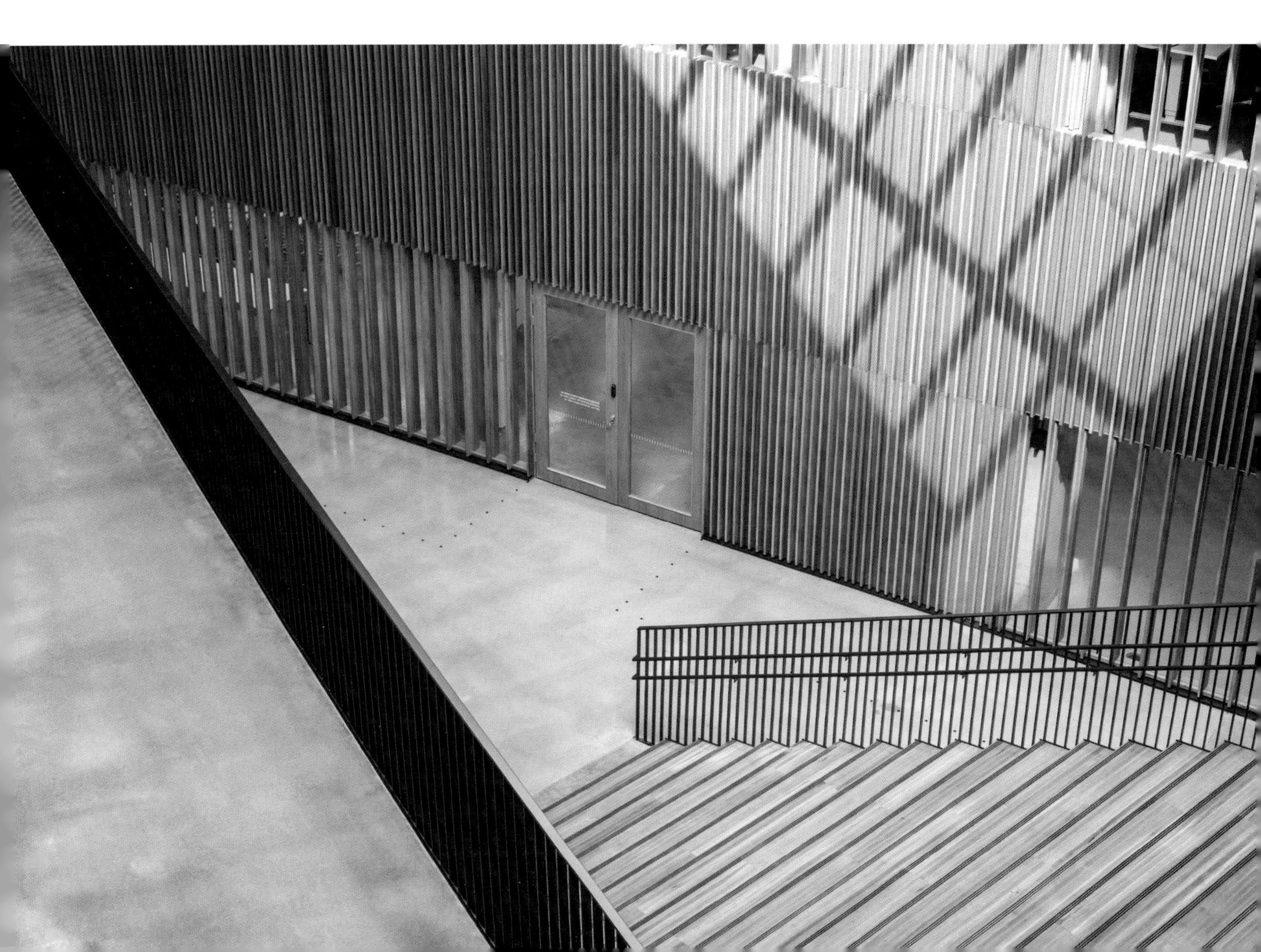

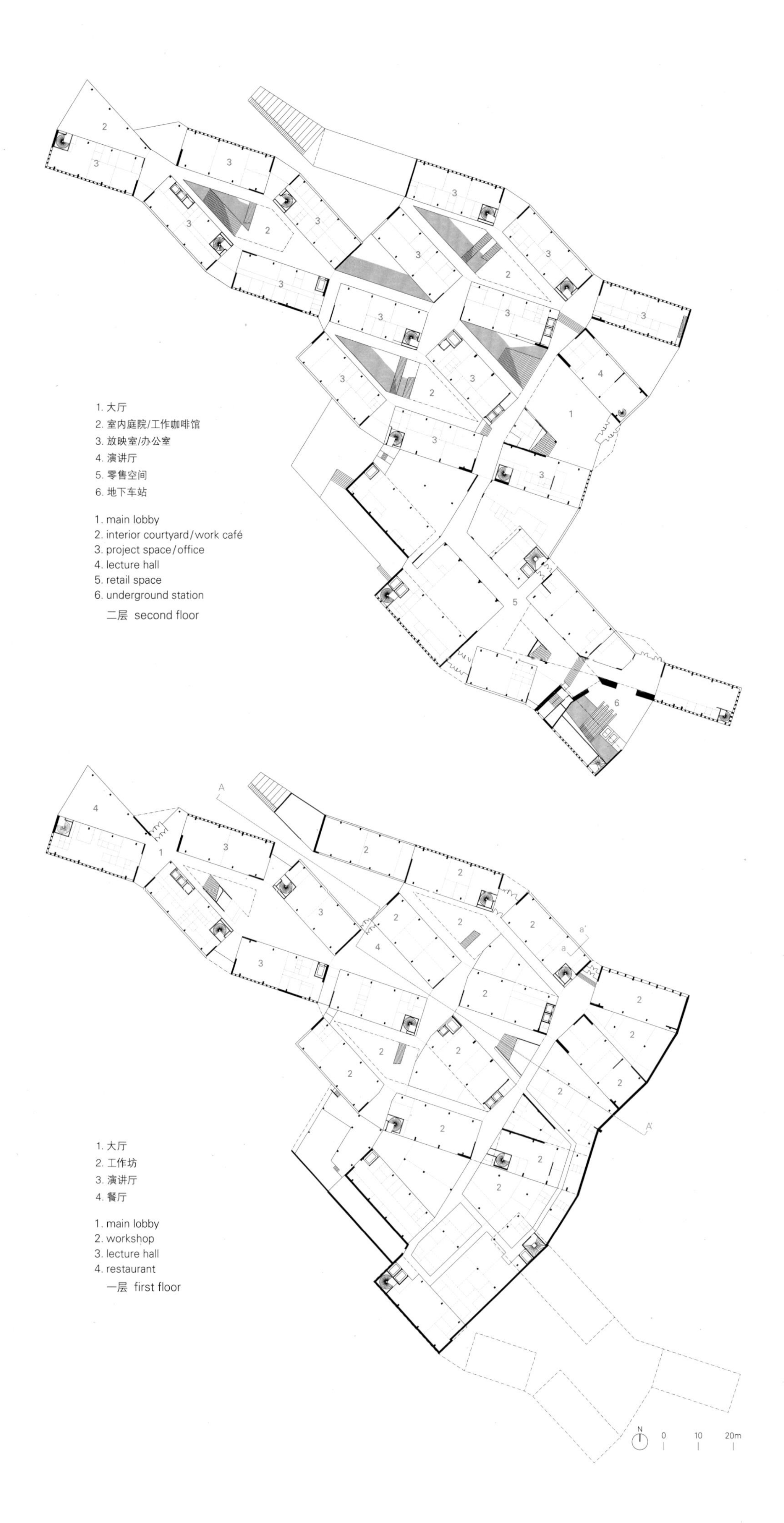
1. 大厅
2. 室内庭院/工作咖啡馆
3. 放映室/办公室
4. 演讲厅
5. 零售空间
6. 地下车站
1. main lobby
2. interior courtyard/work café
3. project space/office
4. lecture hall
5. retail space
6. underground station
二层 second floor
1. 大厅
2. 工作坊
3. 演讲厅
4. 餐厅
1. main lobby
2. workshop
3. lecture hall
4. restaurant
一层 first floor
A
A'
a
a'
N
0
10
20m

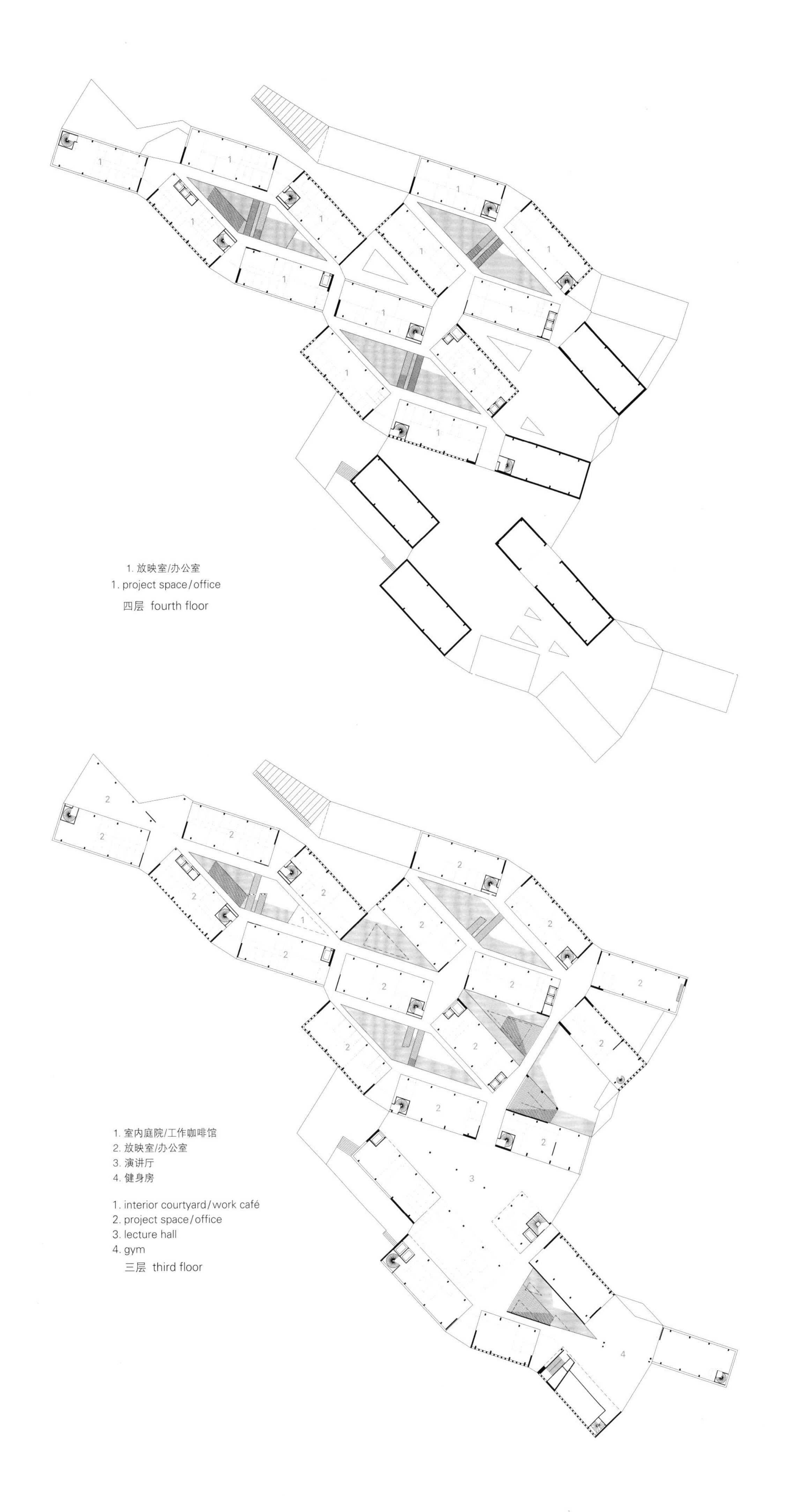
1. 放映室/办公室
1. project space/office
四层 fourth floor
1. 室内庭院/工作咖啡馆
2. 放映室/办公室
3. 演讲厅
4. 健身房
1. interior courtyard/work café
2. project space/office
3. lecture hall
4. gym
三层 third floor

The design forms an entirely new central square that connects the old and new university buildings, an entrance to a new underground line and a shopping center included in the Väre building. The 7,300m^2 shopping center has restaurants, cafés, a gym and retail spaces. Thus, the new square and the Väre Building serves as a place of casual encounters and while it also provides a series of spaces for interdisciplinary interaction for 1,850 students and 350 employees.
The building adapts the classic courtyard university typology to contemporary ways of learning and pedagogy. The spatial layout based on a lattice of simple project space units supports interaction and flexible use. The carefully planned spatial sequence unfolds with rooms arranged around sky-lit atriums that open views to all the floors and the stairs connecting them. Transparent walls reveal the creative action in the project spaces and the ground floor lobby is open to the public. The layout and massing break the large volume of Väre into a cluster of smaller units helping it adapt to the context and scale of the park-like campus. The two principal facade alignments are derived from the old main buildings. The use of hand-laid red brick follows Aalto's masterful treatment of the material while the use of glass patterned glass gives the building a contemporary expression. Inside, wood surfaces highlight the flow of the spatial sequence through the building. The schools of design and business have their own characters. In the School of Arts, Design and Architecture, the sculptural, self-supporting steel stairs stand out against a neutral backdrop. In the School of Business, limestone, oak and a warm metal color create an ambiance that is both cozy and dignified.
The Väre Building is environmentally friendly, energy-efficient and water-saving. 200 bicycle parking places and changing rooms encourage cycling. Ground-source heat pump covers 90% of the heat consumption and roof-top solar panels around 6% of the electricity consumption. Ventilation is demand-controlled and has heat recovery. Lighting is controlled, too. Furthermore, 75% of the construction waste was recycled.

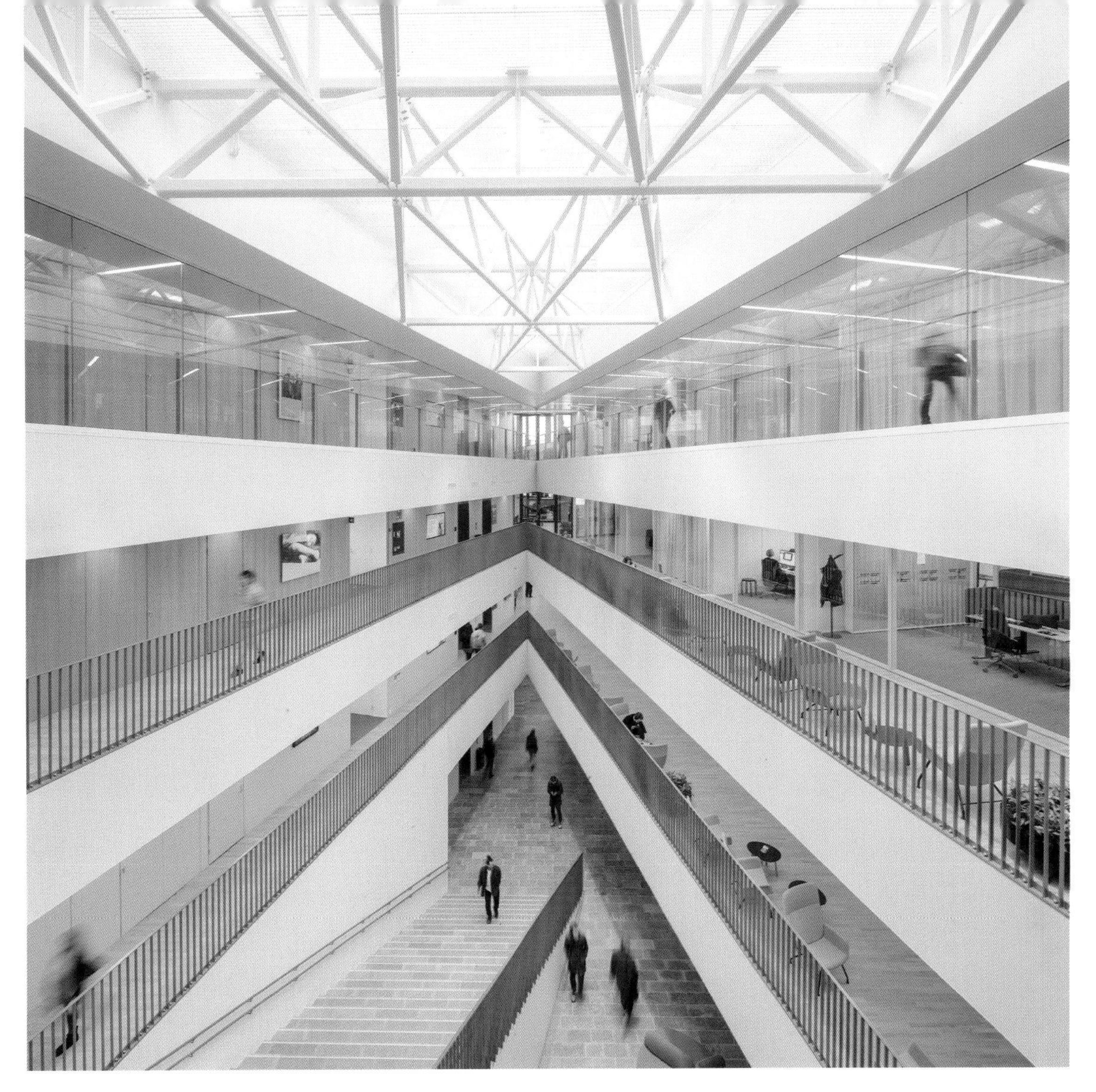

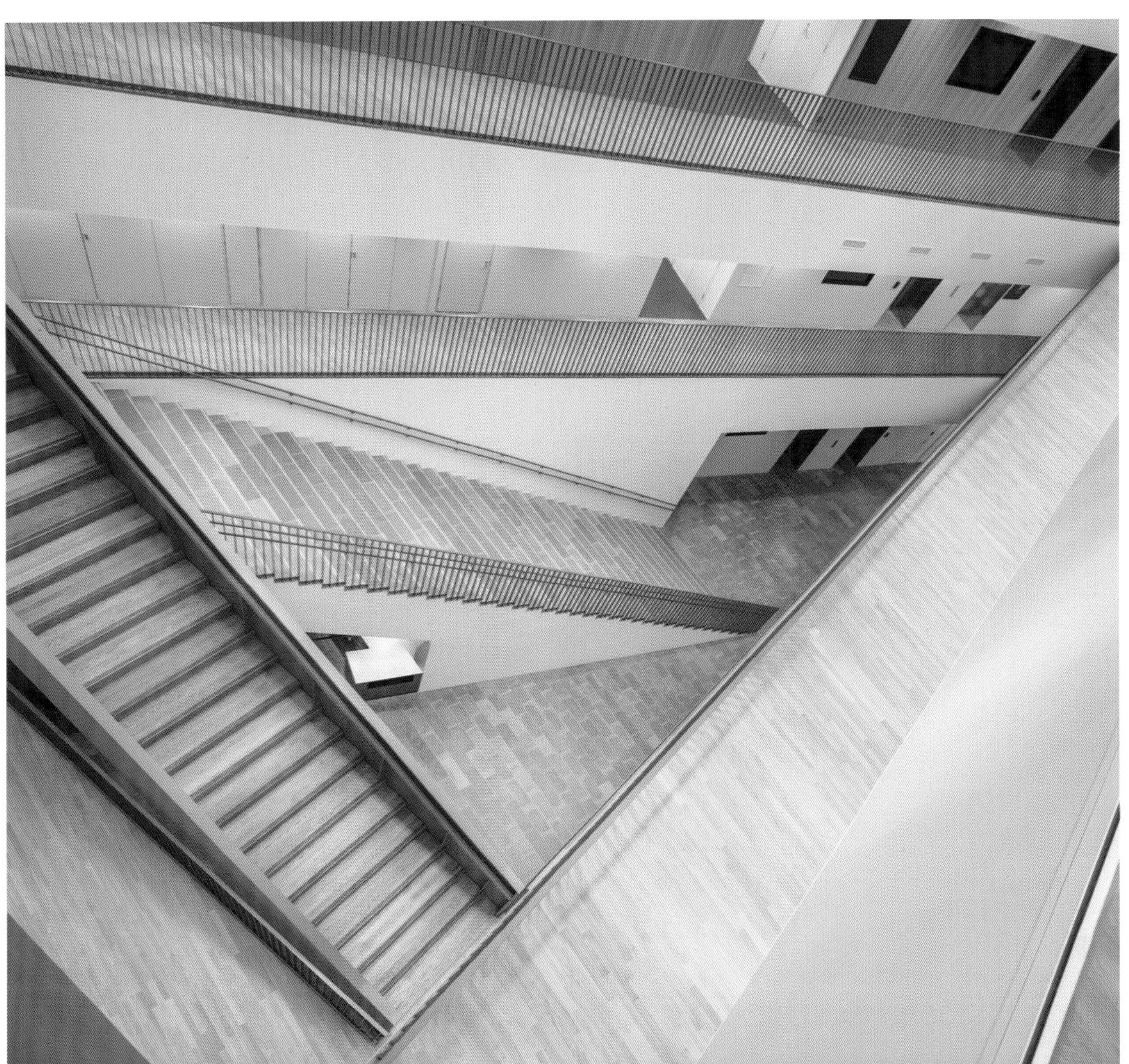

格拉西尔-托尔斯港学院
Glasir Tórshavn College

BIG

格拉西尔学院坐落在法罗群岛令人叹为观止的景色之中
Glasir nestled in the breathtaking scenery of the Faroe Islands

项目位于法罗群岛起伏的峡湾地带，可以望见首府托尔斯港、大海以及树木繁茂的山丘。为了获得更高的效率，格拉西尔学院将法罗群岛体育馆、托尔斯港技术学院和法罗群岛商业学院合并在一栋建筑内，以供1750多名学生、教师和员工使用。

格拉西尔学院分别保留了三所学校的自主性和个性，同时为学校之间的合作以及教学工作的蓬勃发展提供了理想的环境。格拉西尔学院由五个独立的楼层围绕着中央庭院堆叠而成：三所学校各占一层，教师办公区和餐厅占一层，体育锻炼和集会空间则占用了最后一个楼层。建筑的形态犹如一个漩涡，每一层都充分向外打开，其中顶层朝着山丘地带延伸了30m。

考虑到陡峭的地形，在进入学校主入口时需要先通过一座尺度巨大的桥梁。大型的露天圆形庭院立即映入了眼帘，仿佛在向每一位到来的学生和老师张开双臂，为所有的楼层和学术空间都创造了一个自然的集会地点。

格拉西尔学院设计反映了自然景观，直径为32m的室内阶梯式露台为许多不同用途提供了宽敞、灵活的空间，例如，小组会议、社会活动和餐饮，以及大型事件或广播的礼堂座位。项目第一页的图中，巨大的透明天窗抵挡住了法罗群岛大风等恶劣气候的影响，同时也为中庭提供了充足的阳光。教室和庭院之间的内立面采用彩色玻璃，直观地概述了建筑内的不同功能。

阶梯式的地形跨越多个层次将建筑合并为一个整体。位于顶部的高中和商学院通过悬臂式的体量向山丘和旷野打开，使建筑与四面八方的景观融为一体。不论是从庭院、教室，还是从体育馆和图书馆，都可以随时随地感受到法罗群岛的壮观景象。

室内空间选用了石材、模铸混凝土以及表面以不同方式处理的木材，共同营造出干净而自然的气息；外部主要使用了玻璃和铝材，以中性的质感衬托出建筑的独特造型。

从外部看，玻璃材质的立面被镶嵌在呈锯齿状排布的面板之间，使直角的单元结构得以形成一个柔和的圆形体量，从而融入周围的环境。随着时间的推移，屋顶上的绿植将变得更加繁茂，整座建筑也将消隐在法罗群岛的景观当中。

Located on the undulating fjords of Faroe Islands, with views to the capital Tórshavn, the sea and verdant fells, Glasir seeks to harvest the efficiencies of combining Faroe Islands Gymnasium, Tórshavn Technical College and the Business College of Faroe Islands into one building for over 1,750 students, teachers and staff.

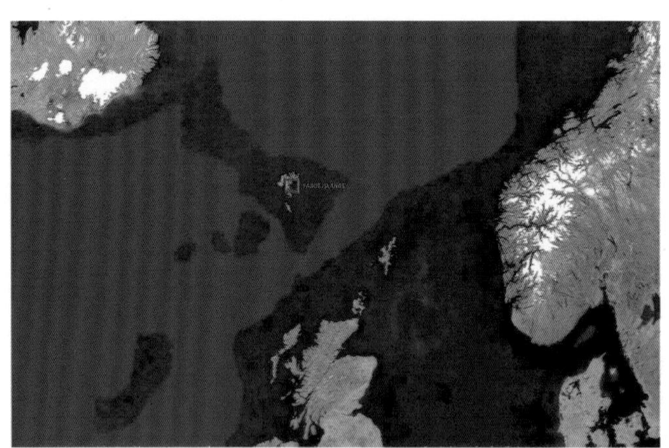

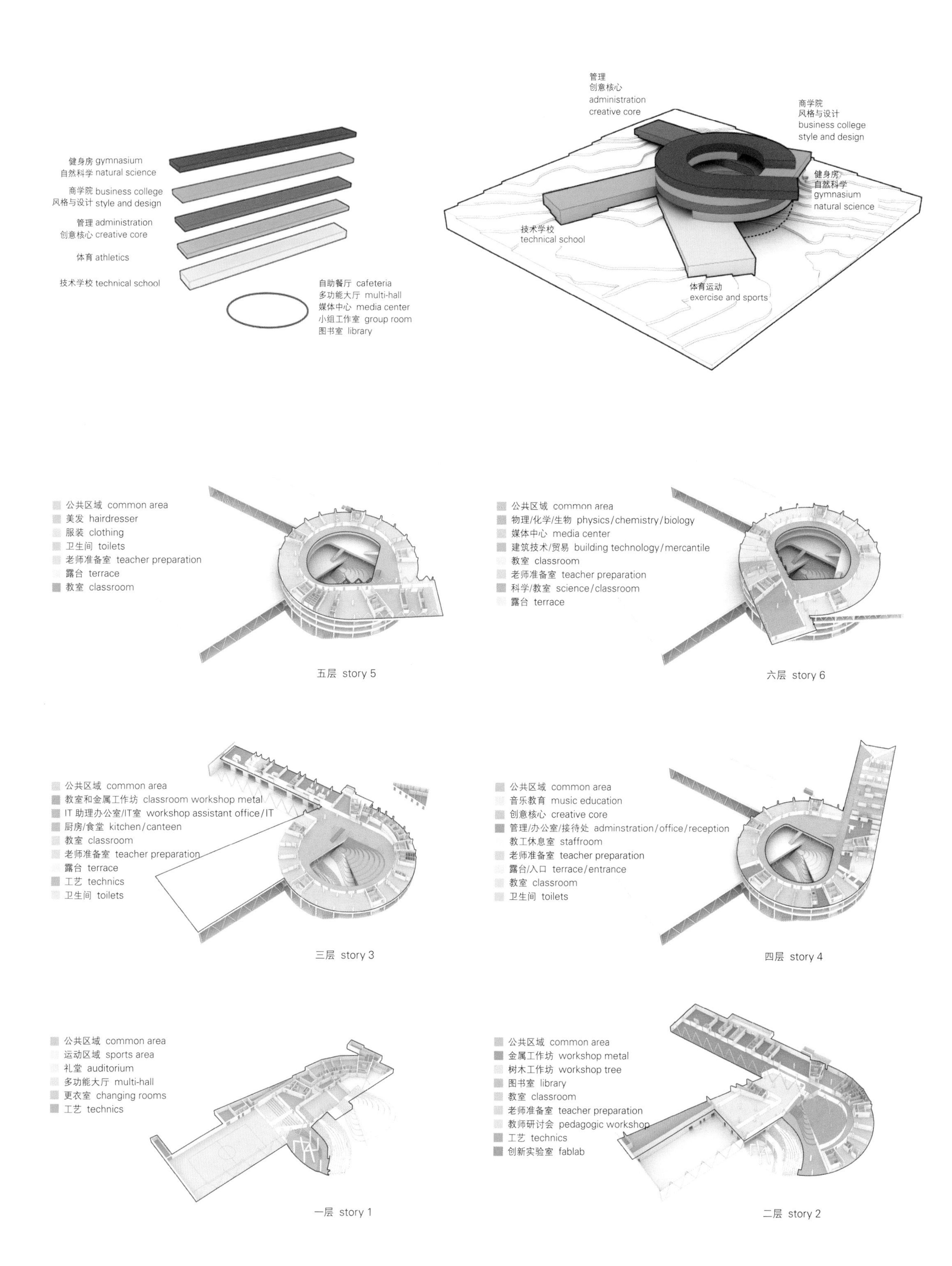

健身房 gymnasium
自然科学 natural science
商学院 business college
风格与设计 style and design
管理 administration
创意核心 creative core
体育 athletics
技术学校 technical school
自助餐厅 cafeteria
多功能大厅 multi-hall
媒体中心 media center
小组工作室 group room
图书室 library
管理
创意核心
administration
creative core
商学院
风格与设计
business college
style and design
健身房
自然科学
gymnasium
natural science
技术学校
technical school
体育运动
exercise and sports
公共区域 common area
美发 hairdresser
服装 clothing
卫生间 toilets
老师准备室 teacher preparation
露台 terrace
教室 classroom
五层 story 5
公共区域 common area
物理/化学/生物 physics/chemistry/biology
媒体中心 media center
建筑技术/贸易 building technology/mercantile
教室 classroom
老师准备室 teacher preparation
科学/教室 science/classroom
露台 terrace
六层 story 6
公共区域 common area
教室和金属工作坊 classroom workshop metal
IT 助理办公室/IT室 workshop assistant office/IT
厨房/食堂 kitchen/canteen
教室 classroom
老师准备室 teacher preparation
露台 terrace
工艺 technics
卫生间 toilets
三层 story 3
公共区域 common area
音乐教育 music education
创意核心 creative core
管理/办公室/接待处 adminstration/office/reception
教工休息室 staffroom
老师准备室 teacher preparation
露台/入口 terrace/entrance
教室 classroom
卫生间 toilets
四层 story 4
公共区域 common area
运动区域 sports area
礼堂 auditorium
多功能大厅 multi-hall
更衣室 changing rooms
工艺 technics
一层 story 1
公共区域 common area
金属工作坊 workshop metal
树木工作坊 workshop tree
图书室 library
教室 classroom
老师准备室 teacher preparation
教师研讨会 pedagogic workshop
工艺 technics
创新实验室 fablab
二层 story 2

Glasir retains the autonomy and individual identity for each of the three schools while creating ideal conditions for collaboration and learning to flourish. Glasir is conceived as a stack of five individual levels that wrap around a central courtyard: one for each of the three institutions, one for food and faculty, and one for physical exercise and gatherings. The building is organized like a vortex, with each level opening up and the top levels radiating 30m / 100ft out towards the mountainous landscape.

The main entrance of the school is accessible from a dramatic bridge due to the steep slope of the site. Students and teachers are immediately welcomed by the large circular courtyard which creates a natural gathering point across all floor levels and academic interests.

Designed as an extension and interpretation of the natural landscape, the 32m diameter indoor courtyard with terraced steps provide generous, flexible spaces for group meetings, social events and dining, as well as auditorium seating for larger events or announcements. Above, gigantic transparent skylights shelter the space from the wind and weather of the harsh Faroese climate while allowing abundant daylight into the atrium. The inner facades between the classrooms and courtyard are realized in colored glass, providing an intuitive overview of the different functions within the building.

Cascading across several levels, the stepped topography merges the multistory building into a single entity. At the top levels, the high school and business school cantilever towards the mountain-range and moorland landscapes, creating a building that open towards the city in all directions. The outer backdrop to the stunning Faroese landscape surrounding the education center is always visible, from the courtyard and classrooms to the gymnasium and library.

The interior material selection of stone, cast concrete and wood with different surface treatments form a natural canvas, while the exterior choice of glass and aluminum creates a neutral background to the school's sculptural design.

Seen from outside, the exterior glass facades are mounted in a sawtooth shingle that allow the straight elements to form a soft circular shape and blend into the natural surroundings. Over time, grass planted on the rooftops will slowly grow to allow the education center to disappear into the Faroese landscape.

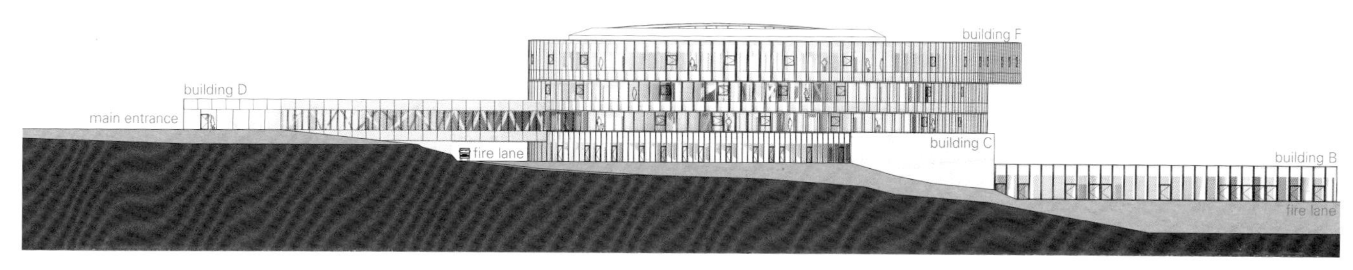

东立面 east elevation

北立面 north elevation

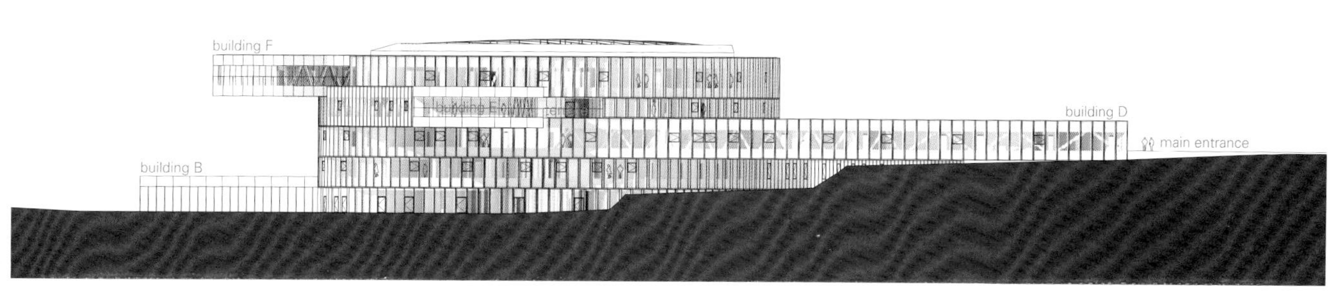

西北立面 north-west elevation

西南立面 south-west elevation

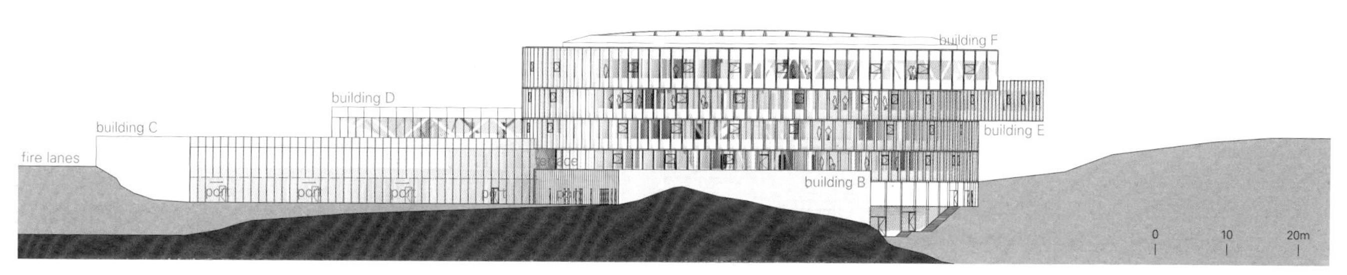

南立面 south elevation

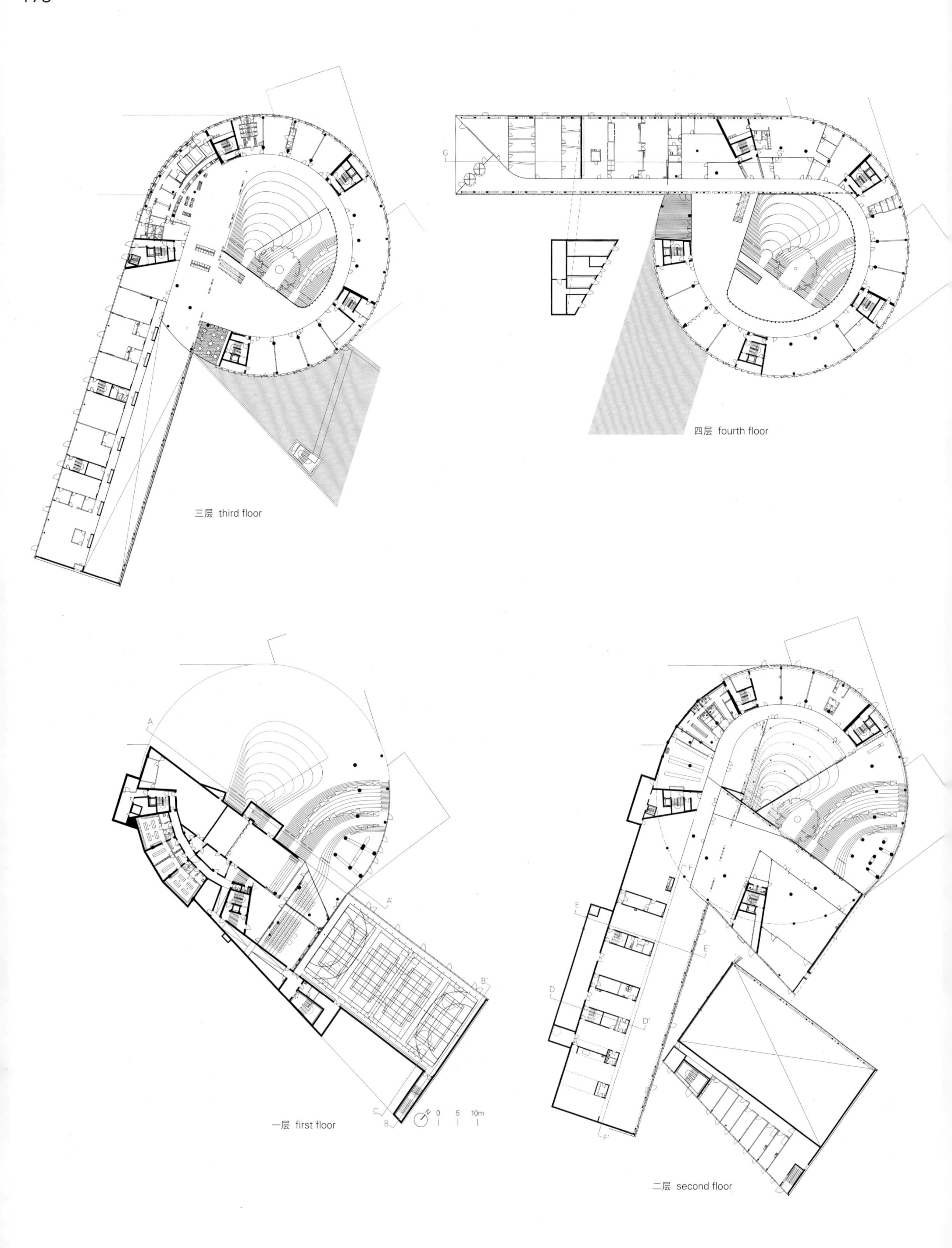

三层 third floor

四层 fourth floor

一层 first floor

二层 second floor

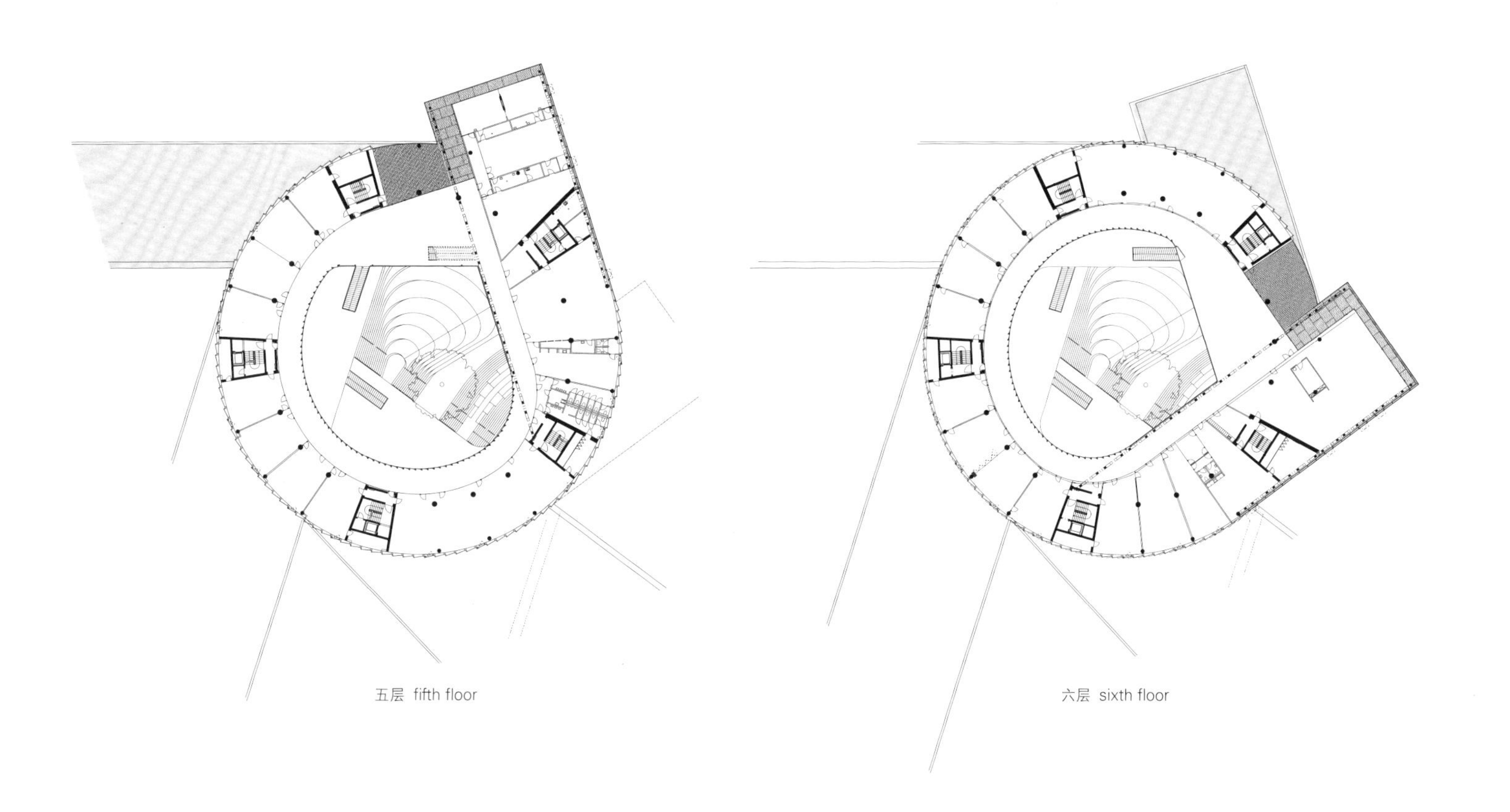

五层 fifth floor

六层 sixth floor

E-Vent DX

项目名称：Glasir Tórshavn College / 地点：Tórshavn, Faroe Islands / 建筑师：BIG / 负责合伙人：Bjarke Ingels, Finn Nørkjær, Ole Elkjær Larsen / 项目建筑师：Høgni Laksáfoss
项目团队：Alberte Danvig, Alejandro Mata Gonzales, Alessio Valmori, Alexandre Carpentier, Annette Birthe Jensen, Armen Menendian, Athena Morella, Baptiste Blot, Boris Peianov, Camille Crepin, Claudio Moretti, Dag Præstegaard, Daniel Pihl, David Zahle, Edouard Boisse, Elisha Nathoo, Enea Michelesio, Eskild Nordbud, Ewelina Moszczynska, Frederik Lyng, Goda Luksaite, Henrik Kania, Jakob Lange, Jakob Teglgård Hansen, Jan Besikov, Jan Kudlicka, Jan Magasanik, Jeppe Ecklon, Jesper Boye Andersen, Ji-Young Yoon, Johan Cool, Kari-Ann Petersen, Kim Christensen, Long Zuo, Martin Cajade, Michael Schønemann Jensen, Mikkel Marcker Stubgaard, Niklas Rausch, Norbert Nadudvari, Oana Simionescu, Richard Howis, Sabine Kokina, Simonas Petrakas, Sofia Sofianou, Takumi Iwasawam, Tobias Hjortdal, Tommy Bjørnstrup, Victor Bejenaru, Xiao Xuan Lu / 创意：Tore Banke, Kristoffer Negendahl / 合作方：Fuglark, Lemming & Eriksson, Rosan Bosch (space planning, furnishing, wayfinding), Samal Johannesen, Martin E. Leo SP/F, KJ Elrad Radgevandi Verkfroendingar / 客户：Mentamalaradid (Ministry of Culture), Landsverk / 用途：education / 总建筑面积：19,200m² / 竣工时间：2018 / 摄影师：©Rasmus Hjortshoj (courtesy of the architect)

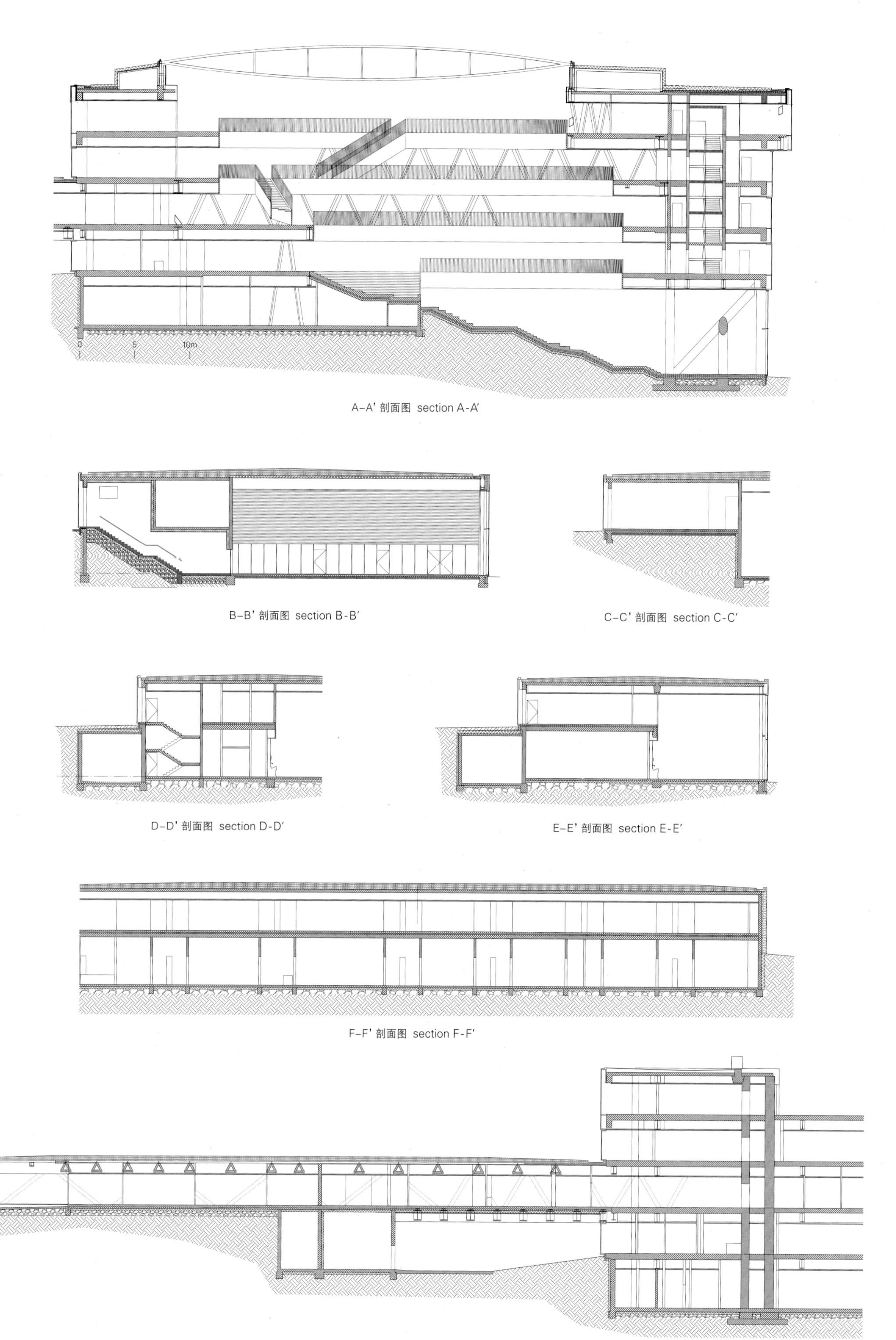

A–A' 剖面图 section A-A'

B–B' 剖面图 section B-B'

C–C' 剖面图 section C-C'

D–D' 剖面图 section D-D'

E–E' 剖面图 section E-E'

F–F' 剖面图 section F-F'

G–G' 剖面图 section G-G'

马士基医学科研大楼
The Maersk Tower

C.F. Møller Architects

C.F. Møller 建筑事务所用一个标志性的新地标扩建了丹麦大学的潘侬区医疗中心

C.F. Møller extends the Panum medical complex at the University of Denmark with an iconic new landmark

马士基大楼是一栋世界顶级的科研大楼，其创新性的建筑为世界尖端医学研究创造出了一个最优化的框架结构建筑，这让它成了哥本哈根的地标性建筑。同时它也让哥本哈根大学与周边的社区和广阔的城市连接了起来。

这座大楼是哥本哈根大学健康与医疗学院的扩建工程，里面包含了科研和教学设施，以及一个带有礼堂和会议室的会议中心，并且都装配了最先进的技术。

为了打造一个具有世界顶尖水平的医学研发大楼，设计一个促使许多机会汇聚在一起、超越不同学科、适合大众及研究界的场所是十分重要的。这样设计有助于就正在进行的研究活动进行交流，实现知识共享，并为新兴的、具有开创性的研究提供灵感。

对于城市绿化和城市公园来说，塔楼式的建筑类型会带来很大的益处，这里对每个人都是开放的，因此它与周边的发展规划都是有一定联系的。新园区公园的一个独特元素是蜿蜒曲折的"飘浮路径"，它引导行人和骑行者穿过马士基大楼的部分区域。这让公众有机会在接近大楼和研究人员的同时，在诺尔街和布莱格丹斯韦街这两条相邻的街道之间建立新的联系。

该大楼位于一个由共享公共设施构成的、低矮的星形基础上方。整个基地拥有透明立面，因此显得开放和热情，同时，这种透明度使得大楼内部融入了外部的绿色景观。

这栋大楼在其创新型和现代化的实验室中拥有所有的研发设施。每层楼的功能都是通过一个有效的循环路径连接在一起的，这样可以缩短行走的距离，并能够加强团队合作的机会。

设计师设计了一个连续的螺旋旋转楼梯，让人们不论从视觉上还是身体上都和这开阔的15个楼层的中庭连接在了一起，让空间呈现出一种宽敞的立体感。每个楼层的楼梯附近都有一个开放的、引人前来的"科学广场"，这是员工们用于自然交流的公共空间。在大楼正面的铜百叶窗上有一块巨大的垂直的玻璃碎片，这样在外面就能够看到里面的螺旋楼梯和科学广场，与开放式的基座一起确保了楼内相关活动的可视性，同时还能让人们欣赏到哥本哈根的美丽壮观而又鼓舞人心的景色。

该大楼的立面被划分成了浮雕状的网格结构，上面覆盖着一层楼高的铜盖百叶窗。这种百叶窗是一个可移动的气候防护罩，可根据阳光直射的情况自动打开或关闭，确保实验室的直接热增益保持在绝对最小值。与此同时，百叶窗也可以为立面提供一种立体浮雕效果，将大楼庞大的表面分解开来。在这些表现中，百叶窗也营造了一种精细、垂直的感觉。

马士基大楼拥有丹麦最节能的实验室所在地，这里的剩余能源循环利用达到了一个前所未有的水平。再结合立面的可移动式热屏蔽系统和其他节能措施，该建筑成了节能实验室建筑的典范。

The Maersk Tower is a state-of-the-art research building whose innovative architecture creates the optimum framework for world-class health research, making it a landmark in Copenhagen. It aims to contribute positively by linking the University of Copenhagen with the surrounding neighborhoods and wider city.

The Tower is an extension of Panum, the University of Copenhagen's Faculty of Health and Medical Sciences. It contains both research and teaching facilities, as well as a conference center with auditoriums and meeting rooms, connected to the latest technology.

In order to create architecture for world-class health research, it is important to design a venue which encourages many opportunities for coming together, transcending different disciplines, and suitable for the general public and the research community. This helps to communicate ongoing research activities, leading to knowledge sharing and providing inspiration for new and groundbreaking research.

塔楼部分 tower level

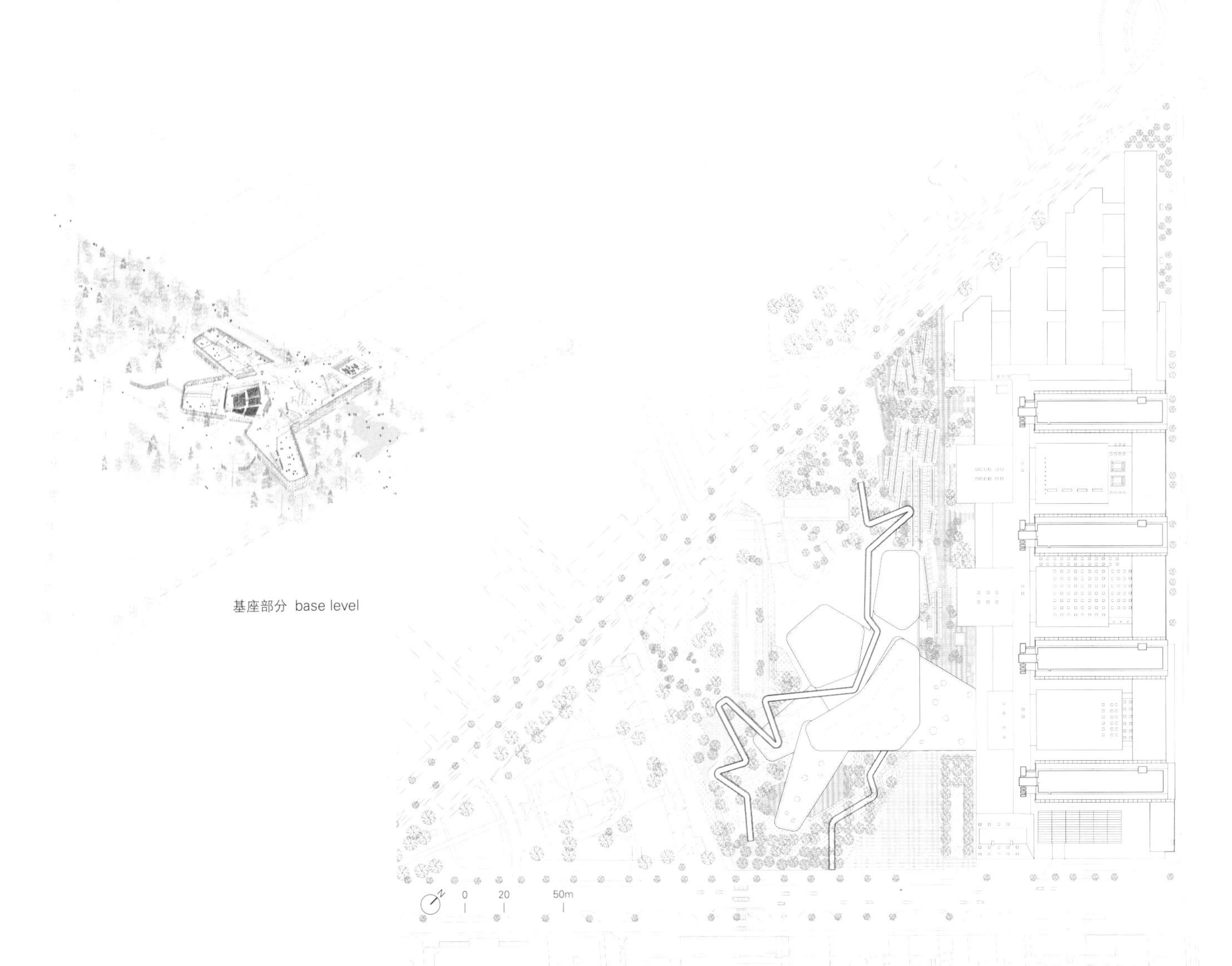

基座部分 base level

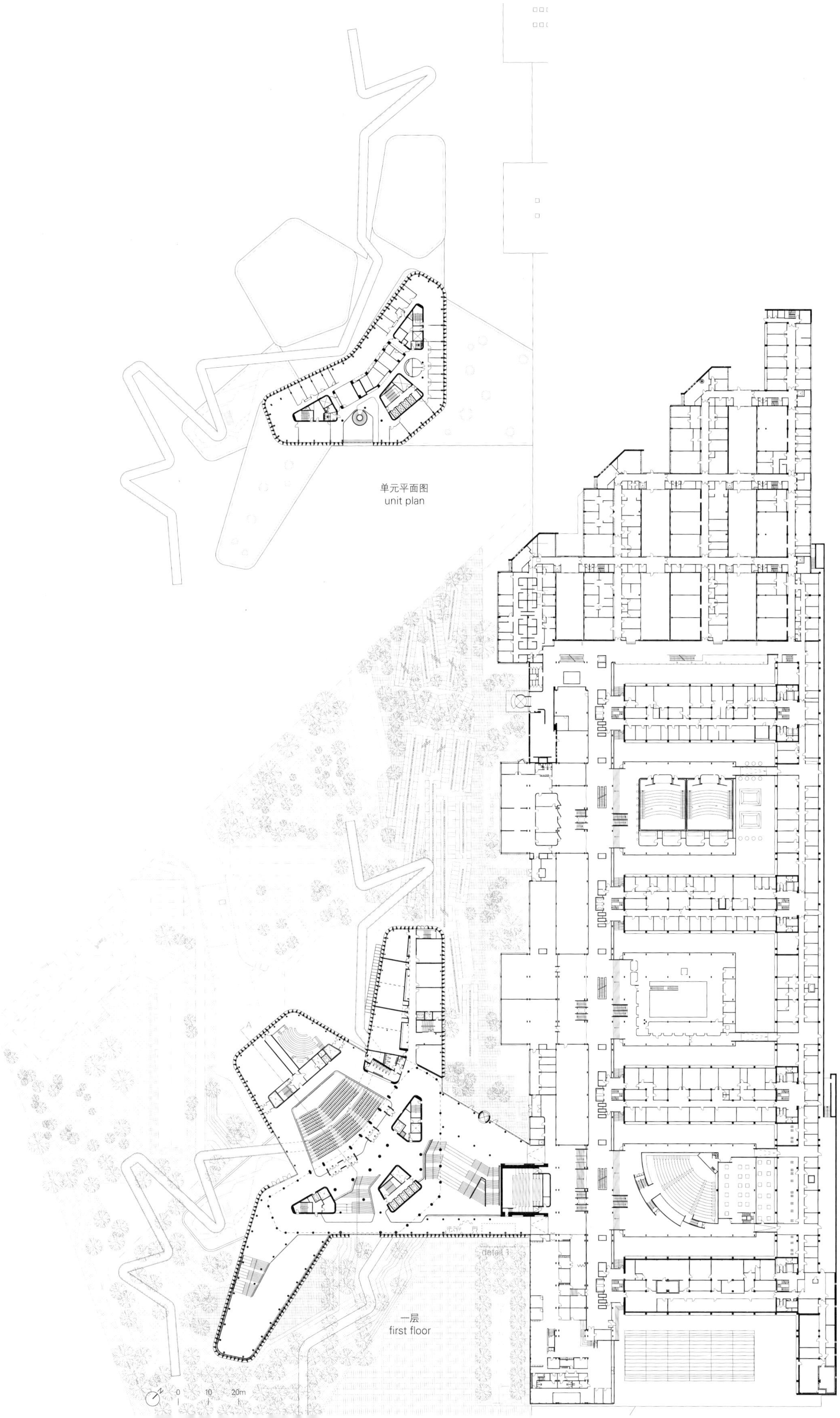

单元平面图
unit plan
一层
first floor
detail 1
0
10
20m

By selecting a tower typology, there was greater allowance for a green and urban campus park, which is open to everyone and therefore involves and develops the surrounding neighborhood. A unique element of the new Campus Park is the zigzagging "floating path" that leads pedestrians and cyclists across parts of the Maersk Tower. This allows the public the opportunity to get up close to the building and the researchers while, at the same time, creating a new connection between Nørre Allé and Blegdamsvej, the adjacent streets.
The tower rests on a low, star-shaped base which contains shared and public facilities. With its transparent facade, the entire base appears open and welcoming and at the same time this transparency allows the interior of the building to blend in with the surrounding green landscape.
The tower itself holds all research facilities, in innovative and modern laboratories. On each floor the tower's functions are linked together in an efficient loop, which provides shorter travel distances and strengthens opportunities for teamwork. A continuous sculptural spiral staircase visually and physically connects the open fifteen-floor atrium, creating an extensive three-dimensional sense of space. Close to the staircase on each floor there is an open and inviting "Science Plaza", which serves as a natural meeting and communal space for the many employees. A large vertical shard of glass, in the copper shutters of the facade, makes the spiral staircase and the science plazas visible externally, and ensures, together with the open base, visibility in relation to the activities of the tower as well as a spectacular and inspiring view over Copenhagen.
The facade of the Tower is divided into a relief-like grid structure of story-height copper-covered shutters. The shutters of the facade function as movable climate shields, that automatically open or close according to direct sunlight, ensuring that direct heat gain in the laboratories is kept to an absolute minimum. At the same time, the shutters provide a deep relief effect to the facade, breaking down the considerable scale of the tower. In their expression, the shutters also offer a sense of fineness and verticality.
The Maersk Tower hosts Denmark's most energy-efficient laboratories, where waste energy is recycled to an unprecedented level. This, in combination with the movable heat-shielding of the facade and other energy-saving measures, makes the building a pioneer of energy-efficient laboratory construction.

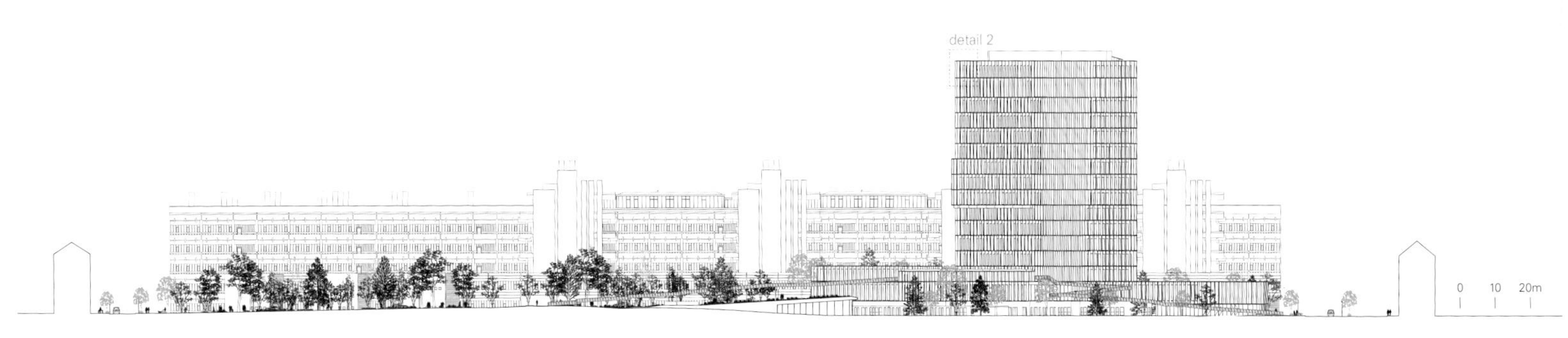

西南立面 south-west elevation

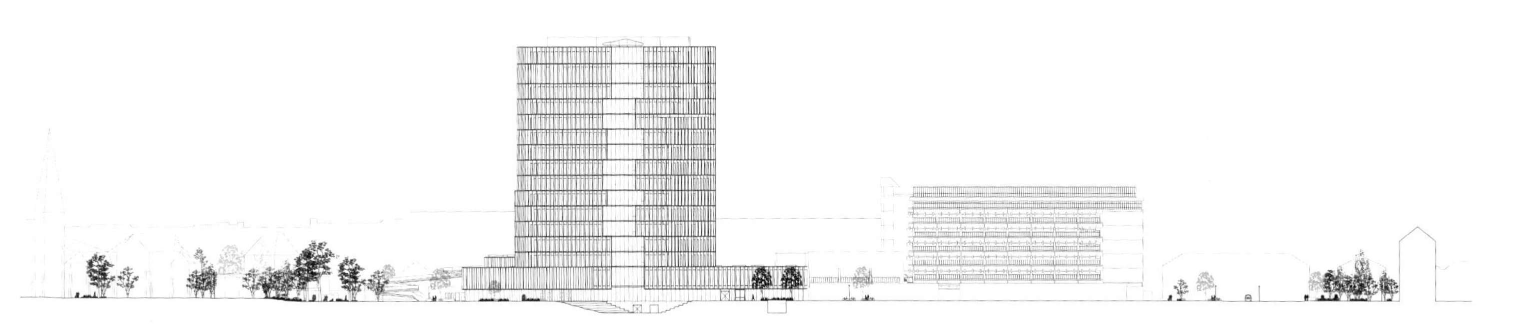
东南立面 south-east elevation

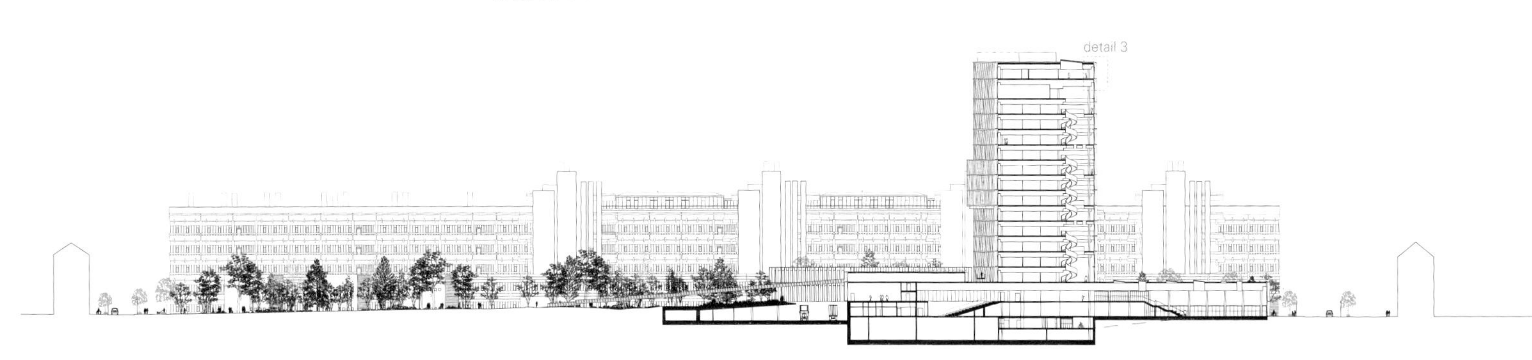

A-A' 剖面图 section A-A'

项目名称：The Maersk Tower / 地点：Nørre Campus, Blegdamsvej, Copenhagen, Denmark / 建筑师：C.F. Møller Architects / 景观设计师：SLA / 工程师：Rambøll
合作方：Aggebo & Henriksen, Cenergia, Gordon Farquharson, Innovation Lab / 客户：The Danish Property Agency for the University of Copenhagen supported by the A.P. Møller Foundation / 总建筑面积：42,700m² (24,700m² laboratories, offices and shared facilities and 18,000m² foyer, canteen, auditoria, classrooms, plant) / 竣工时间：2017
摄影师：©Adam Mørk (courtesy of the architect), except as noted

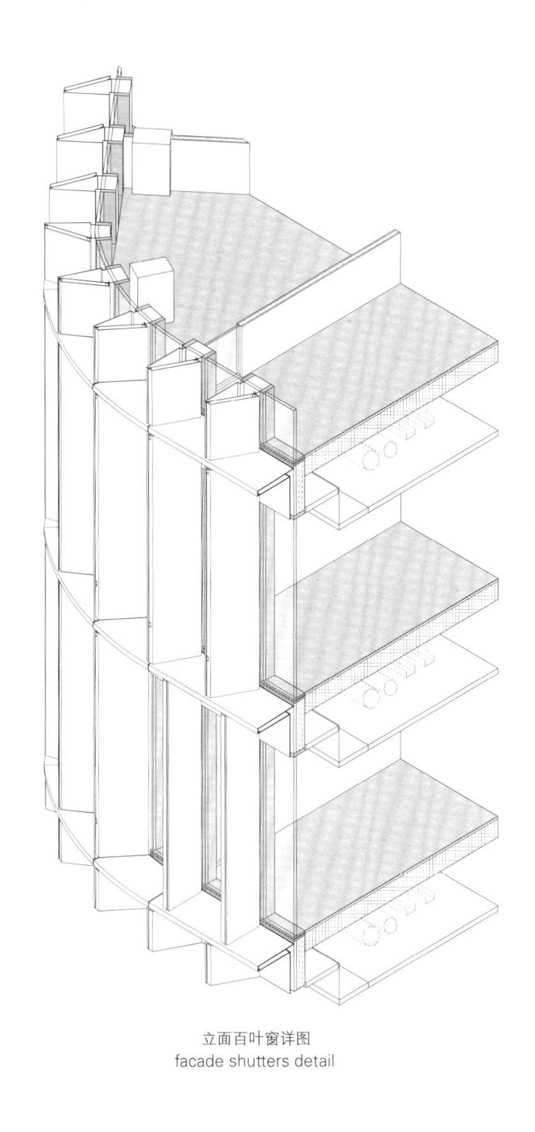

立面百叶窗详图
facade shutters detail

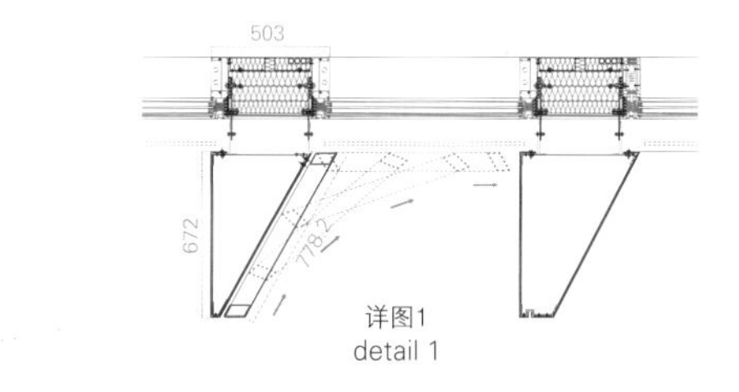

详图1
detail 1

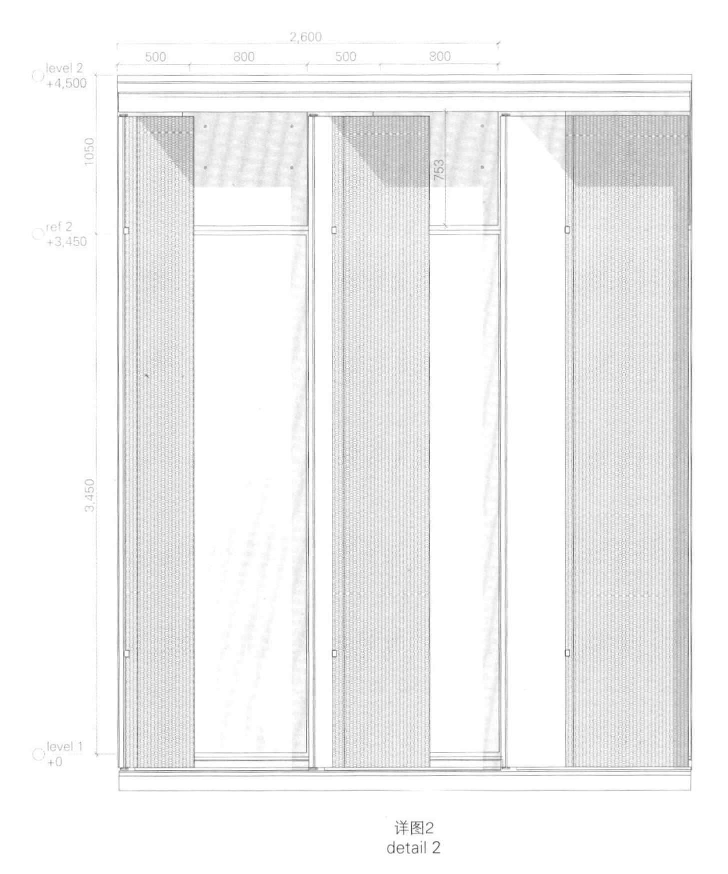

详图2
detail 2

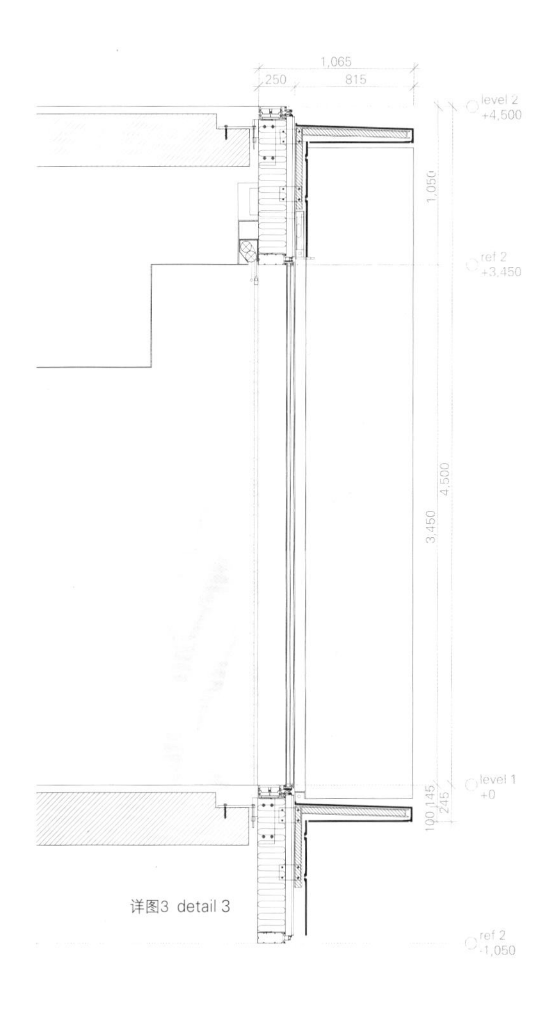

详图3 detail 3

Vandrehallen
Auditorier
Undervisning

P172 Verstas Architects

Is a Helsinki-based award-winning architectural practice founded by Ilkka Salminen, Riina Palva, Väinö Nikkilä, and Jussi Palva[from left] in 2004. Try to consider various perspectives to produce designs that are human-centered and believe this to be the most sustainable approach to architecture that withstands time. Their team consists of approximately 40 architects, urban designers, interior and landscape architects who are able to work at various scales from masterplanning down to intricate details. The name Verstas (Finnish for workshop) describes their methodology which entails close-knit work between clients and the project team to ensure a tailored final design.

P108 Civic Architects

Specializes in public buildings, libraries, bridges, theatres, concert halls, art galleries, museums, schools and residential complexes. Its design respects and anticipates social, cultural, economic and ecological requirements. Currently has a staff of 10 architects led by 4 managing partners. The result is functional and aesthetically responsible architecture which serve the needs of generations to come. (picture below: Ingrid van der Heijden, Rick ten Doeschate, Jan Lebbink, Gert Kwekkeboom, from left)

Sanjay Mohe

P72 Mindspace

Graduated from Sir J.J. College of Architecture, Sanjay Mohe has been the founding partner of Mindspace in Bangalore for the past 13 years. Has worked in association with Bangalore-based prestigious architectural firm, Chandavarkar & Thacker for 21 years. Also has had a short stint with Charles Correa in Mumbai and his work stint in Saudi Arabia. His work spans a spectrum of Projects-Research Laboratories, Knowledge Parks, Institutions, Resorts, Libraries, Corporate Offices, Hospice and Residences. A well-known face in a lot of Architectural forums and talks, he has a lot of awards to his credit. Received Great Master's Award from J.K. Cements in 2019.

P154 Perkins and Will

An interdisciplinary, research-based architecture and design firm, was founded in 1935 on the belief that design has the power to transform lives. Guided by its core values – design excellence, diversity and inclusion, resilience, social purpose, sustainability, and well-being – the firm is committed to designing a better, more beautiful world. With an international team of more than 2,700 professionals, the firm has over 20 studios worldwide, providing services in architecture, interior design, branded environments, urban design, and landscape architecture.

P16 3XN

Is a Copenhagen-based studio with more than 30 years experience. At the core of its design philosophy is the principle that architecture, if done right, can shape behavior. This happens through the careful analysis of the site, the surrounding buildings, and the client aspirations. Among 3XN's high profile projects are 'The Blue Planet (National Aquarium) (2013)', 'Royal Arena (2017)' in Copenhagen, 'Museum of Liverpool (2011)', 'Quay Quarter Tower (2020)' and the 'New Sydney Fish Market (2023)' in Sydney.

P34 Behnisch Architekten

Was founded in 1989 and works out of 3 offices – Stuttgart, Munich, and Boston. These offices are directed by Stefan Behnisch and his partners Robert Hösle (Munich), Robert Matthew Noblett (Boston), Stefan Rappold and Jörg Usinger (Stuttgart). Has a global reputation for high-quality architecture that integrates environmental responsibility, creativity, and public purpose. The firm produces a rich variety of buildings mainly in Europe and North America. The five partners and staff share a vision to push the boundaries of high performance, 21st-century architecture that respects user needs, ecological resources, and local cultures.

P208 C.F. Møller Architects

Is one of Scandinavia's oldest and largest architectural practices. Was founded in 1924 by the now deceased Danish architect C. F. Møller. Today it has more than 350 employees in Head office in Aarhus, Denmark and some branches in Copenhagen, Aalborg, Oslo, Stockholm and London. Over the years, have won a large number of national and international competitions. Also, has been exhibited locally as well as internationally at places like at RIBA in London, the Venice Biennale, and the Danish Cultural Institute in Beijing. In 2015, it was named one of the Top 10 Most Innovative Companies in Architecture.

Klaus Toustrup, Lone Wiggers, Mads Mandrup Hansen, Christian Dahle, Michael Kruse, Lone Bendorff, Mårten Leringe, Klavs Hyttel, Julian Weyer, from left.

P142 **ZAmpone Architectuur**

Founded in 2007, is a Brussels based architectural office with Tom De Fraine, Bart Van Leeuw and Karel Petermans[from left] as partners. Is a multidisciplinary team consisting of 12 people who have own specialization and distinct individuality. They have specialized in socially relevant construction projects in an urban and, more specifically, a Brussels context. The enormous variety of programs and the open exploratory attitude of them ensure a great variety within the projects. Without pinning themselves on a preconceived architecture concept, they look for the most suitable solution that provides an inventive answer to the question and the budget, both architecturally, constructively and sustainably, as well as from the point of origin of the program and organization.

P190 **BIG**

Founded in 2005 by Bjarke Ingels, BIG is a Copenhagen, New York and London based group of architects, designers, urbanists, landscape professionals, interior and product designers, researchers, and inventors. Currently involves in a large number of projects throughout Europe, North America, Asia, and the Middle East. Believes that in order to deal with today's challenges, architecture can profitably move into a field that has been largely unexplored. A pragmatic utopian architecture that steers clear of the petrifying pragmatism of boring boxes and the naïve utopian ideas of digital formalism. Like a form of programmatic alchemist, it creates architecture by mixing conventional ingredients such as living, leisure, working, parking, and shopping. By hitting the fertile overlap between pragmatic and utopia, once again finds the freedom to change the surface of our planet, to better fit contemporary life forms.

P92 **Vurpas Architectes**

Vurpas Architectes' ultimate, ongoing ambition is to create places for people. Pays close attention to the use of materials, natural light and color to build spaces which meet contemporary requirements in terms of comfort and well-being. Benefits from its passion for regenerating buildings and heritage sites, and from its extensive experience and expertise to create a conversation between the project and its setting. In line with their shared values, Julien Leclercq, Daniel Briet, Philippe Beaujon, Brigitte Scharff, and Damien Pontet[from left] continue the development of the firm initiated by Pierre Vurpas.

P4 **Bart Lootsma**

Is a historian, theoretician, critic and curator in the fields of architecture, design and the visual arts. Is a Professor for Architectural Theory and Head of the Institute for Architectural Theory, History and Heritage Preservation at the University of Innsbruck. Was Guest Professor for Architecture, European Urbanity and Globalization at the University of Luxemburg; at the Academy of Visual Arts in Vienna; at the Akademie der Bildenden Künste in Nürnberg; at the University of Applied Arts in Vienna; at the Berlage Institute in Rotterdam and Head of Scientific Research at the ETH Zürich, Studio Basel. Was the guest curator of ArchiLab 2004 in Orléans. Was an editor of *ao*. Published numerous articles in magazines and books.

P8 **Herbert Wright**

Writes about architecture, urbanism and art. Graduated in Physics and Astrophysics from the University of London. Is contributing editor of UK architecture magazine *Blueprint*, and has contributed to publications including *Wired*, *RIBA Journal*, *The Guardian*, *l'Architecture d'Aujourdhui*, *Abitare* and *C3*. His books include *London High* (2006) and *Instant Cities* (2008). Curated Lisbon's first city-wide public architecture weekend Open House 2012, was short-listed to curate Oslo Architecture Triennial 2016, was a 2017 Graduate Thesis juror at SCI-Arc, and keynote speaker at Element Urban Talks, Krakow 2018.

P86 **Heidi Saarinen**

Is a designer, educator, architectural writer and dance artist with specialism in concepts, theories and experimental methodologies. Her continued research investigates peripheral places, thresholds and the interaction/collision between architecture and the body. Is working on a series of collaborative projects connecting interiors, architecture, memory and place. Is part of several community and creative groups in London where she engages in events and projects highlighting awareness of community and architectural conservation in the built environment.

P166 **Francesco Zuddas**

Is Senior Lecturer in Architecture at Anglia Ruskin University. Previously taught at Università degli Studi di Cagliari, the Architectural Association, Central Saint Martins and the Leeds School of Architecture. Writings on postwar Italian urbanism and architecture, the spatial implications of changing production paradigms towards the knowledge economy, have appeared on *AA Files*, *Domus*, *Territorio*, and *Trans*, among other publications. Is the co-author of *Territori della Conoscenza: Un Progetto per Cagliari e la sua Università* (2017). and of *Made in Taiwan: Architecture and Urbanism in the Innovation Economy* (2012). Latest book is *The University as a Settlement Principle: Territorialising Knowledge in Late 1960s Italy* (2019).

P52 **ERA Architects**

Ali Hiziroglu[picture right-bottom] graduated from Ecole Spéciale d'Architecture: Paris with honorable mention in 2001, where he also worked as an assistant in architecture studios. He completed his master's degree in architecture and urban design in Columbia University, NYC. He began to work for ERA Architects in 2003, in 2010 became a partner and managed the design phases of large-scale mixed-use, office projects. World Architecture Festival (WAF) finalist with Garanti BBVA Technology Campus in 2019, Shenyang Starmall Plaza in 2009. He continues as a partner designer to work in the field of architecture and urban design at ERA Architects' Istanbul, Beijing and Moscow offices.

François Jean Victor Valentiny

P126 **Valentiny hvp Architects**

François Jean Victor Valentiny[picture-above] studied Architecture at Nancy University and received his M.Arch at University of Applied Arts Vienna in 1980. After his graduation, he opened 'Hermann et Valentiny' with Hubert Hermann. Is a professor at the Beijing DeTao Masters Academy in Shanghai since 2011. Changed the name of 'Hermann Valentiny et Associés' to 'Valentiny hvp Architects' in 2012. Founded 'Valentiny Foundation' in 2014 and was selected as one of 100 architects of the year 2016 by the Korean Institute of Architects & UIA. Is currently being led by François Valentiny and 4 directors & associates including Axel Christmann, Jeanne Petesch, Laurye Pexoto and Daniela Flor.

Ali Hiziroglu

图书在版编目(CIP)数据

历史价值的适应性再利用 / 丹麦BIG建筑事务所等编；司炳月，许正阳，姜博文译. — 大连 ：大连理工大学出版社，2020.12

ISBN 978-7-5685-2691-3

I . ①历… II . ①丹… ②司… ③许… ④姜… III . ①建筑设计 IV . ①TU2

中国版本图书馆CIP数据核字(2020)第176265号

出版发行：大连理工大学出版社

（地址：大连市软件园路80号　　邮编：116023）

印　　刷：上海锦良印刷厂有限公司

幅面尺寸：225mm×300mm

印　　张：14.75

出版时间：2020年12月第1版

印刷时间：2020年12月第1次印刷

出 版 人：金英伟

统　　筹：房　磊

责任编辑：张昕焱

封面设计：王志峰

责任校对：杨　丹

书　　号：978-7-5685-2691-3

定　　价：298.00元

发　行：0411-84708842

传　真：0411-84701466

E-mail：12282980@qq.com

URL：http://dutp.dlut.edu.cn